Synthesis Lectures on Digital Circuits & Systems

Series Editor

Mitchell A. Thornton, Southern Methodist University, Dallas, USA

This series includes titles of interest to students, professionals, and researchers in the area of design and analysis of digital circuits and systems. Each Lecture is self-contained and focuses on the background information required to understand the subject matter and practical case studies that illustrate applications. The format of a Lecture is structured such that each will be devoted to a specific topic in digital circuits and systems rather than a larger overview of several topics such as that found in a comprehensive handbook. The Lectures cover both well-established areas as well as newly developed or emerging material in digital circuits and systems design and analysis.

Selahattin Sayil

Noise Contamination
in Nanoscale VLSI Circuits

 Springer

Selahattin Sayil
Department of Electrical Engineering
Lamar University
Beaumont, TX, USA

ISSN 1932-3166   ISSN 1932-3174   (electronic)
Synthesis Lectures on Digital Circuits & Systems
ISBN 978-3-031-12753-3   ISBN 978-3-031-12751-9   (eBook)
https://doi.org/10.1007/978-3-031-12751-9

This Springer imprint is published by the registered company Springer Nature Switzerland AG
The registered company address is: Gewerbestrasse 11, 6330 Cham, Switzerland

To,
-My Wife
Neziha Sayil
and
-My Sons
Kaan Sayil and Kutay Sayil

Preface

As CMOS technology advances, circuits are exposed to various noise sources that impact circuit performance and reliability in nanometerscale integrated circuits. The reliability of the integrated circuits is being severely challenged, as many previously negligible noise effects are becoming more prominent, causing significant performance and reliability degradations of nanometer integrated circuits.

Various noise sources exist in a nanometer circuit including crosstalk noise induced by signal switching on neighboring wires, power supply noise on power grid lines due to simultaneous switching events, substrate coupling noise, noise due to process and environmental variations, thermal noise and radiation-induced soft errors. The impacts of these noise sources are far more severe for clock lines since clock signal controls the data processing.

Increasing circuit frequencies and reducing power supply voltages sub-1V make reliability of circuits more challenging due to reduced noise margins of gates. As the number of devices placed on a single chip increase exponentially with the integration, and with GHz range operation frequencies, the stronger and more frequent interactions among neighboring transistors occur leading to stronger noise effects and higher failure rate.

Unlike other textbooks that just cover a single noise mechanism, this textbook covers various noise sources and the relationship between these noise sources and compound effects is also explored. In addition, modeling and mitigation of some of the important noise mechanisms have also been discussed.

Beaumont, USA   Selahattin Sayil

Contents

About the Author

Selahattin Sayil received the M.Sc. degree from the Pennsylvania State University, University Park, PA, in 1996 and the Ph.D. degree in Electrical Engineering from Vanderbilt University, TN, in 2000. He is currently a Professor and Distinguished Faculty Fellow in Electrical Engineering at Lamar University, where he leads the VLSI CAD and Signal Integrity Group.

His current research interests include mitigation of radiation effects in VLSI, interconnect delay and noise analysis, low-power design, and testing. His teaching interests include online teaching, high-impact online laboratories, and virtual team learning. He has authored two books, two book chapters, and published nearly thirty refereed journal articles that include special feature articles, and presented at many international conferences. He is a member of IEEE and serves as an Associate Editor for *International Journal of Electronics*.

The word "noise" in digital circuits refers to any occurrence that may cause the voltage at a node to deviate from its nominal value which can be either a nominal supply or ground. The magnitude of noise is characterized by the peak noise amplitude with respect to the nominal value. Although digital circuits have more noise immunity in comparison to analog circuits, noise immunity in these circuits has become equally important to area, timing and power in newer technologies due to scaling effects.

As the feature sizes continue to scale further in deep nanometer scale, various noise mechanisms are becoming stronger than ever [1]. Rising clock frequencies in the GHz region and reducing supply voltages in the sub-1 V range are further exacerbating the impact of these mechanisms.

With the scaling of today's nanoscale CMOS technologies, integrated circuits operate with lower power supply voltage, higher clock frequency, and higher device and interconnect density. The high performance microprocessors now include billions of transistors and recent technologies such as Intel's 14 nm process [2] can easily include 13 layers of interconnect. All these factors lead to an incredibly noisy on-chip environment as illustrated in Fig. 1.1.

Owing to different physical mechanisms, many noise sources co-exist in a nanometer circuit. These include crosstalk noise induced by signal switching on neighboring wires, power supply noise or IR voltage drops on power grid lines due to simultaneous switching events of buffers/gates, substrate coupling noise, noise due to process and environmental variations during the chip's operation and radiation induced noise effects.

The supply voltage levels keep getting closer to the threshold level meaning that the amount of noise buffer we have to protect signals from these perturbances are getting smaller and smaller. With the decreased noise margin of transistors, nanometer circuits become increasingly more vulnerable to noise interferences.

© The Author(s), under exclusive license to Springer Nature Switzerland AG 2022   1
S. Sayil, *Noise Contamination in Nanoscale VLSI Circuits*, Synthesis Lectures on Digital Circuits & Systems, https://doi.org/10.1007/978-3-031-12751-9_1

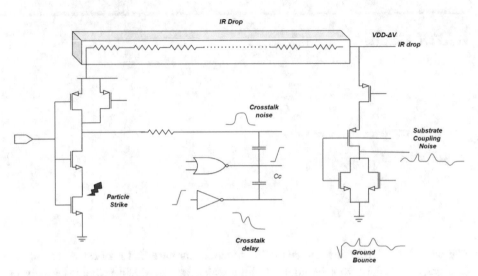

Fig. 1.1 Various noise sources in nanoscale VLSI

With advancements in technology scaling, on-chip crosstalk noise between interconnects has become a major signal integrity issue [3]. The reduced dimensions of wires along with shortened distances and high wire aspect ratios give rise to increased coupling capacitances.

Increased circuit frequencies and various layers of interconnect make interconnect coupling noise even more important in newer technologies. With coupling, a signal transition occurring on a wire now can transfer some of its charge to its neighbor wires results in voltage glitches or perturbances. The severity of these glitches also increases as the signal transitions get sharper due to faster waveforms. Higher circuit frequencies also increase the likelihood of these glitches corrupt the signal on neighboring wires more frequently and more efficiently.

In digital integrated circuits, capacitive and inductive coupling can create significant noise effects interfering with the circuit's regular operation. The glitch effect occurs when the affecting wire also known as the aggressor switches and then couples noise to a neighboring wire that is non-switching. If the coupled wire is also in switching state, the coupling effect now can cause an increase or a decrease in signal transition time and may affect timing of the neighboring circuit. In other words, the crosstalk noise can introduce uncertainty to the propagation delay of a circuit. In this case, the coupling capacitance between neighboring interconnects varies with different signal switching scenarios, the crosstalk hence introduces uncertainty to the propagation delay of a circuit.

Power supply noise is another critical issue in nanoscale circuit design [4]. Smaller device dimensions, increasing circuit frequencies, and lower power supply voltage all make digital VLSI systems more vulnerable to power supply noise. The chip power distribution systems are normally designed to provide constant supply voltage across a chip reliably over its lifetime. However, due to increased metal line resistance, package pin inductance, higher current levels and due to changes in current demand, power (V_{DD}) and ground distribution networks (V_{SS} lines) may fluctuate. The voltages on these lines are no longer constant and are impacted by two different types of voltage drops: IR voltage drops and Ldi/dt voltage variations.

The ohmic IR drops on power and ground distribution networks occur due to variations of the average current demand which is input pattern dependent. With increasing number of devices, the average current demand in an IC increases in newer technologies. Larger chip sizes also increase the line resistance of the power and ground distribution networks due to increased distances thereby worsening the impact of resistive power and ground noise. The IR voltage drops occurring on the ground line is usually referred as the ground bounce. Whether it is the IR drop or the ground bounce, both types of the IR noise degrade noise margin of logic gates. As a result, the impacted gates suffer performance losses and overall timing and functionality will be affected.

The effect of IR voltage drops are worsened by the Ldi/dt voltage variations which can occur at package pins. Increase in the frequency raises the rate of change in the current in deep submicron technologies. As a result, Ldi/dt voltage drops become more pronounced in higher frequencies.

Dynamic circuits are also more susceptible to noise disturbance because during some part of their operation the evaluation nodes are disconnected from power and ground. In static circuits momentary deviation of logic levels can be restored automatically, since at steady state the nodes are always connected either to ground or V_{DD}. However, this restoration is not possible in dynamic circuits due to the possibility of floating nodes.

Leakage noise in digital circuits mainly comes from sub threshold conduction due to the off current of FET devices which allows charge to drain from or accumulate on the dynamic node. Another leakage noise source is minority carrier back-injection into substrate due to bootstrapping.

These currents can either charge or discharge a dynamic node or cause the stable state of a weakly held node to be significantly different from rails. Leakage current is most critical, since in digital circuits it exponentially increases with the continuous shrinking of the MOS transistor dimensions.

Charge sharing noise is produced due to the charge redistribution between a weakly held dynamic node and internal nodes of the circuit. Charge sharing noise is caused by charge redistribution between the dynamic node and the internal nodes of the pull down network. Charge sharing reduces the voltage level at the dynamic node causing potential false switching of a dynamic logic gates. As coupling becomes stronger, noise injected into a dynamic circuit via coupling increases.

One of the most critical signals in a synchronous digital circuit is the clock signal. The design of clock distribution networks in synchronous digital systems presents enormous challenges [5]. Controlling the clock signal delay in the presence of various noise sources, process parameter variations, and environmental effects represents a fundamental problem in the design of high-speed synchronous circuits. It is important to reduce the uncertainty of the clock signal delay, particularly of the clock signals driving the registers belonging to the most critical data paths. The delay uncertainty introduced by interconnect coupling in signal lines other than clock signals also imposes additional constraint on the length of clock period.

The transient errors created by terrestrial radiation can also cause increasing reliability issues and compromise the security in unpredictable ways. The radiation-induced soft errors in commercial nanometer CMOS technologies have become a growing concern [6, 7].

While soft errors in memory have been a very well-studied problem at terrestrial level, soft errors are now increasingly affecting CMOS logic due to increasing clock frequencies and diminishing device sizes [8]. For 45 nm technologies and below, researchers has predicted that the majority of the observed radiation induced soft errors will be due to transients that will occur in combinational logic (CL) circuits.

As scaling of today's process technologies continues, circuits become increasingly become more vulnerable to radiation-induced soft errors in nanoscale CMOS technologies. The reduced node capacitances, supply voltages together with increasingly denser chips are raising soft error rates and making them an important design issue. Increasing clock frequencies also increase circuit vulnerabilities to these transients as there is now more chance to capture these transients.

At ground level, soft errors are mainly induced by alpha particles emitted from trace radioactive impurities in the device materials, interaction of low-energy thermal neutrons with certain boron isotopes in the device, and reaction of high-energy cosmic neutrons (>1 MeV) with silicon and other device materials. Neutron induced soft errors are generated by secondary charged particles which are created in neutron-silicon atom collisions. Since a neutron does not carry any charge, it does not induce ionization by itself in silicon. When high energy neutrons strike an integrated circuit, some neutrons pass through without affecting operations of the semiconductor device, but some neutrons collide with nuclei in the silicon lattice. The result of this interaction is the creation of secondary particles, which in turn create a trail of electron–hole pairs. Although alpha particles are directly ionizing, the energy of the secondary ions produced by high-energy neutron reactions can be much higher than that of alpha particles [7, 8].

The alpha particles are emitted mostly due to radioactive decay of uranium and thorium impurities located within the chip packaging. Linear Energy Transfer (LET) is the measure of energy that is transferred in the material when an ionizing particle passes through it. Most alpha particles have energies of between 3 and 7 MeV (mega or million electron-volts). Since it takes only about 3.6 eV to generate an electron–hole pair in the substrate, a 4 MeV alpha particle striking the sensitive node within a combinational logic (CL) can generate more than a million electron hole pairs within its particle track due to ionization

Fig. 1.2 Transient current pulse generation due to particle hit on sensitive node

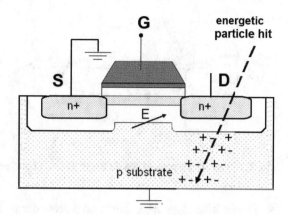

mechanism (Fig. 1.2). The sensitive areas mentioned here are usually the depletion regions of transistor drains or reverse-biased p-n junctions [9]. These would include the drain/well and drain/substrate junctions in CMOS transistors.

Under the electric field, these free carriers can drift creating a transient current pulse. The current later results in a charge collection at the struck electric node. It has been shown that the transient current consists of a fast drift and funneling component and a slower contribution from charge diffusion in the silicon substrate.

The amount of the charge collected at a particular node (Q_{col}), depends on various parameters such as device size, bias conditions, doping level, characteristics of the particle hit, and its trajectory [9]. The collected charge results in a voltage transient at the struck node. This transient, also named as Single Event Transient (SET), can travel through a series of logic gates and finally may reach to a storage element under certain conditions. If the generated pulse arrives at the storage element during its latching window, incorrect data can be stored resulting in soft error or a Single Event Upset (SEU).

When the collected charge Q_{col} at a given node exceeds the critical charge Q_{crit} of that node, the generated SET then can propagate and may reach to storage elements under certain conditions. In its simplest term, critical charge of a node is defined as $Q_{crit} = C_{node} * V_{DD}/2$, where C_{node} is the node capacitance. It is usually assumed that a glitch amount reaching half the power supply voltage can propagate through the receiver gate.

There are three masking effects that can prevent soft error generation: electrical, logical and temporal masking effects. In order for an SET to end up in a soft error:

- The transient pulse generated should have sufficient amplitude and width such that it propagates along the succeeding gates without significant attenuation. Hence, electrical masking should not be present.
- The logic path the pulse takes should be enabled by logic inputs. In another words, there should not be any logical masking.

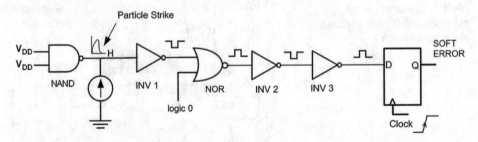

Fig. 1.3 The propagation of Single Event Transient and generation of soft error

- The latching clock edge should be present during the presence of the SET pulse at the input of the storage element. This means no temporal masking should exist.

Figure 1.3 shows all these criteria have been satisfied: i.e. first a sufficient transient pulse is generated at particle site such that it propagates through many stages without any attenuation. There is no logical masking as the second input of NOR gate is tied to logic 0. The pulse arrives during latching edge of the clock pulse; finally soft error is generated due to absence of temporal masking.

Unfortunately, all these masking effects are gradually diminishing with newer generation technologies. With scaling down of devices, the node capacitances reduce, and as a result, electrical masking effects are less due to reduced critical charge. In newer designs, the SET pulses are very comparable to logic pulses. Logical masking is less effective as the logic depth in CL reduces in newer technologies. Temporal masking also reduces as increasing clock frequencies increase the chance of a latching edge being present for registering the data. Increasing circuit frequencies and reducing power supply voltages sub 1 V makes these various noise contributions even more important than ever due to reduced noise margins of gates. As the number of devices placed on a single die increase exponentially with the integration capability, and chips operate at GHz range frequencies, the stronger and more frequent interactions among neighboring devices occur which lead to stronger error effects and higher failure rate.

In mixed IC design, complex digital circuits are often integrated along with high-performance analog circuits on the same substrate [10]. The switching activity generated by digital circuits introduces noise into the common substrate. The substrate becomes the medium that permits noise to easily propagate from digital parts into analog parts within the chip. Analog circuits do not have the wide noise margins of digital circuits and are susceptible to substrate voltage fluctuations. In this case, switching events generated by the digital circuits can interfere with sensitive analog components. When current is injected into the substrate, a fluctuation in substrate voltage occurs and this is usually referred as the substrate noise [10, 11].

On-chip temperature gradient has also emerged as a major design concern for high-performance integrated circuits for the future technology nodes as it significantly affects the performance of a chip. Since the driving strengths of the clock buffers are impacted along with interconnect, clock skew phenomenon for synchronous digital circuits is further elevated by the thermal gradients [12]. In addition, increasing circuit temperature deteriorates circuit reliability and increases circuit sensitivity to soft error and crosstalk events [13].

Finally, process variations caused by the inability to precisely control the fabrication process in newer technologies introduces systematic and random errors into the circuit [14].

The remaining chapters focus on various noise sources that have been just discussed. The chapters will also explore the relationship between these noise sources and compound effects in newer generation technologies.

References

1. International Technology Roadmap for Semiconductors, 2013 Edition, Semiconductor Industry Association (SIA), San Jose, CA.
2. K. Fischer et al., "Low-k interconnect stack with multi-layer air gap and tri-metal-insulator-metal capacitors for 14 nm high volume manufacturing," *IEEE International Interconnect Technology Conference*, 2015, pp. 5–8, doi: https://doi.org/10.1109/IITC-MAM.2015.7325600.
3. K. Bhattacharya and N. Ranganathan, "A unified gate sizing formulation for optimizing soft error rate, cross-talk noise and power under process variations," *10th International Symposium on Quality Electronic Design*, 2009, pp. 388–393, doi: https://doi.org/10.1109/ISQED.2009.481 0326.
4. N. Kanekawa, E. H. Ibe, T. Suga and Y. Uematsu, *"Power Integrity"* in *Dependability in Electronic Systems*, New York, NY, USA: Springer, pp. 91–142, 2011.
5. JM Rabaey, AP Chandrakasan, B Nikolić, *Digital integrated circuits: a design perspective, 2nd ed.*, Pearson Education, 2003.
6. S. Mitra, T. Karnik, N. Seifert, and M. Zhang, "Logic soft errors in sub-65 nm technologies design and CAD challenges," in *Proc. DAC*, pp. 2–3, June 2005.
7. R. D. Schrimpf and D. M. Fleetwood, *Radiation Effects and Soft Errors in Integrated Circuits and Electronic Devices*, World Scientific, 2004.
8. R. C. Baumann, "Radiation-induced soft errors in advanced semiconductor technologies," *IEEE Tran. On Device and Mat. Rel.*, vol. 5, no. 3. pp. 305–316, 2005.
9. P.E. Dodd, L.W. Massengill, "Basic mechanisms and modeling of single-event upset in digital microelectronics," *IEEE Transactions on Nuclear Science*, vol. 50, Issue 3, pp. 583–602, June 2003.
10. Charbon, E., Phillips, J. "Substrate Noise: Analysis, Models, and Optimization", in *VLSI: Systems on a Chip. IFIP*, vol 34. Springer, Boston, MA, 2000, https://doi.org/10.1007/978-0-387-35498-9_41.
11. S. Donnay and G. Gielen, *Substrate Noise Coupling in Mixed-Signal ASICs*, Kluwer Academic Publishers, Boston, MA, 2003.

12. T. Ragheb, A. Ricketts, M. Mondal, S. Kirolos, G. M. Links, V. Narayanan, and Y. Massoud, "Design of thermally robust clock trees using dynamically adaptive clock buffers," *IEEE Trans. Circuits and Systems I*, vol. 56, no. 2, pp. 374–383, Feb. 2009.
13. S. Sayil, P. Bhowmik, "Mitigating the thermally induced single event crosstalk", *Analog Integrated Circuits and Signal Processing*, vol. 92, no. 2, pp. 247–253, 2017.
14. S. Bhunia, S. Mukhopadhyay, K. Roy, "Process Variations and Process-Tolerant Design," *Proc. of Intl. Conf. on VLSI Design*, 2007, pp. 699–704.

Interconnect Noise

2

2.1   Introduction

With the continuous scaling of interconnect dimensions and increasing operating circuit frequencies, the on-chip crosstalk has become a serious reliability issue. As the number of devices per unit area increases, interconnects are more tightly coupled to each other resulting in a higher probability of unwanted crosstalk interference between them.

The scaling process results in a reduction in wire width and spacing in order to save circuit area. Meanwhile, there is a tendency to scale wire thickness (T) slowly compared to the width of the wire (W) to get a better control of the wire resistance. This practice results in tall wires having aspect ratios much greater than unity (T/W >> 1) [1], and these neighboring wires effectively act like a parallel plate capacitor as shown in Fig. 2.1. The reduced dimensions of wires, shortened distances among wires and high wire aspect ratios all give rise to increased cross-coupling capacitances.

With the continuous scaling of chip technologies, coupling capacitance among wires has become the dominant contributor to the total wire capacitance. The increased circuit frequencies and many layers of interconnect make interconnect coupling noise even more prevalent in newer technologies. Although, inductive coupling can create noise impact in digital integrated circuits, capacitive coupling is the most dominant coupling mechanism [2]. Hence, this chapter considers the capacitive coupling mechanism only.

The interaction caused by parasitic coupling between wires is generally known as "crosstalk" and can cause voltage transients or timing changes on neighboring wires. The crosstalk noise or the glitch impact occurs when the switching wire, also known as the "aggressor", couples its noise to quiet neighbors that is in non-switching state. A quiet wire is also named as the "victim wire". In this case, the signal transition occurring on the aggressor wire transfers some of its charge to its neighboring wire resulting in a voltage glitch or a "crosstalk noise".

© The Author(s), under exclusive license to Springer Nature Switzerland AG 2022   9
S. Sayil, *Noise Contamination in Nanoscale VLSI Circuits*, Synthesis Lectures on Digital Circuits & Systems, https://doi.org/10.1007/978-3-031-12751-9_2

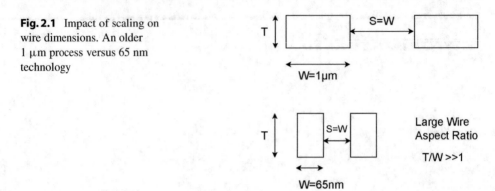

Fig. 2.1 Impact of scaling on wire dimensions. An older 1 μm process versus 65 nm technology

As an example, Fig. 2.2 shows a scenario where an aggressor net switches in rising fashion while victim line is in steady or quiet state. This results in a glitch waveform generation on the victim line in positive direction as shown. The amplitude and duration of this positive glitch or crosstalk noise depends on many circuit parameters such as wire parasitic, drive strength, coupling capacitance, and loads etc.

If the glitch waveform carries sufficient amplitude and width, it can pass through the following receiver gates without any attenuation and may cause a false switching at the register element. In the example shown, the positive glitch generated on victim line due to a rising aggressor causes the switching of the victim receiver (inverter) eventually. With the correct timing present at the latch element, it then causes wrong data to become stored at the output of the latch element.

Associated with a victim line receiver, there exists a noise rejection curve where the failure region is defined. The noise rejection curve defines the amplitude-width combination that results in an error (see Fig. 2.3). A crosstalk glitch having a large magnitude needs to possess sufficient width in order to pass through a receiver gate. Otherwise, it may get filtered and pose no problem at all.

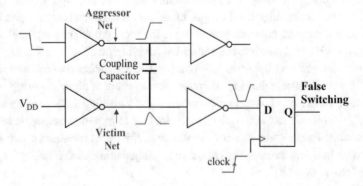

Fig. 2.2 Generation of positive crosstalk pulse

Fig. 2.3 Noise rejection curve

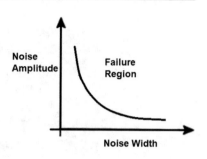

The amount of noise coupled on victim wire depends on wire parasitic, aggressor transition time and victim driver strength. As the circuit frequencies increase, the amount of coupled noise also increases due to rapid transition time of aggressors. Higher circuit frequencies also increase the chance of these glitches corrupting the signal on victim lines.

In the scenario illustrated in Fig. 2.2, if the victim line were to switch simultaneously along with the aggressor, then the coupling effect would have an impact on circuit timing. Thus, the crosstalk noise may introduce uncertainty in the propagation delay of a circuit. Depending on the victim switching direction, it could either increase or decrease the victim signal delay. If the aggressor line switches in the opposite direction to the victim net, this would cause an increase in victim delay, and if the aggressor switches in the same direction to victim line but faster, then this would result in a victim signal switching sooner or a speed-up.

Figure 2.4 shows a scenario where the victim line delay is increased due to opposite switching on the neighboring aggressor net. Here, the aggressor line switches in rising fashion, and couples a positive charge to victim net. Since victim net is in falling state during the noise coupling, the coupled charge results in victim waveform to switch a little later resulting in a signal delay as shown. These delays on victim line can later translate into setup time violations of storage elements.

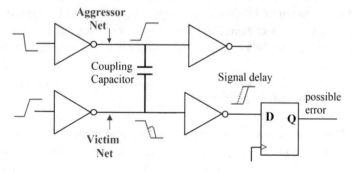

Fig. 2.4 Generation of crosstalk induced delay

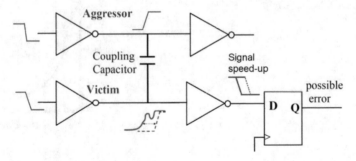

Fig. 2.5 Signal speedup effect due to coupling

The delay impact can clearly be seen at the output of the victim receiver. If it reaches to the input of a register element during/near the clock edge, it can cause timing violations at the register element possibly resulting in incorrect data storage. The switching aggressor's timing makes a huge difference in the amount of delay that it may cause. The maximum delay is achieved when aggressor switches close the 50% of the noiseless victim waveform. Referring to the circuit shown in Fig. 2.4, the aggressor should start switching right before victim waveform falls to the $0.5 * V_{DD}$ which is also known as the threshold of the receiver gate. In addition, the faster the switching on the aggressor line, the larger delay it would impose on the victim.

If the aggressor line switches in the same direction as the victim net, the victim line signal may experience a delay reduction rather than a delay effect. This effect is also known as the "signal speedup" effect. The aggressor line must not only switch in the same direction as the victim line, but it must also switch faster than the victim line for the speedup effect to occur. Otherwise, the coupling capacitor has no charge to transfer and the speedup effect won't occur.

Figure 2.5 shows a case where the speedup impact occurs. As a result of positive charge coupling, the victim line waveform switches sooner than it would. The impact can clearly be seen at the output of victim receiver(s). The decrease in interconnect delay due to coupling impact can affect circuit performance as the speedup effects may later violate hold time requirements of the logic storage elements that are connected to these receivers.

2.2   Miller Factor Analysis

In order to determine the impact of crosstalk on the signal propagation delay, researchers have focused on techniques to directly decouple multiple lines into an equivalent isolated line by using a Miller factor or a Switch factor (SF) [3–5]. Depending on the switching activities on the neighboring wire, the coupling capacitance is converted into an equivalent Miller capacitance that is connected to ground from each node, thus decoupling the victim line from the aggressor for timing calculation purposes. Once decoupled, the Elmore delay formulation can be used to calculate the delay of the isolated wire. The derived worst-case/best-case Miller factors can be used in resolving set-up time/hold time violations of storage elements.

The decoupling Miller factors are very easy to use and integrate into an existing static timing analysis. Some noise-aware static timing engines use switching windows to determine if noise is important. Most of these engines utilize worst-case and best-case Miller factors to find an initial solution [6, 7]. In timing analysis, traditionally Miller factors were taken as either 0 or 2 depending on the switching activity on neighbor line. For example, a ground capacitance of twice the coupling capacitor is taken if signals switch simultaneously in opposite direction (worst-case estimation). When signals switch in the same direction the equivalent capacitance was assumed as zero corresponding to best case delay estimation.

Miller Factor Analysis replaces the coupling capacitance between two neighboring wires with an equivalent capacitance to ground for each line to make it easier to calculate delays using isolated wires. We can use the simple model shown in Fig. 2.6 to illustrate Miller factor calculation. Here, the aggressor and victim nodes are shown by "A" and "V" respectively.

Using the Kirchhoff's Current Law (KCL) at victim node, one can write:

$$C_{eq}(V)\frac{dV_V}{dt} = C_C\frac{dV_V}{dt} - C_C\frac{dV_A}{dt} \qquad (2.1)$$

$$C_{eq}(V) = C_C\left[1 - \frac{dV_A/dt}{dV_V/dt}\right] \qquad (2.2)$$

If V_A and V_B switches in the same direction simultaneously with $R_A = R_V$ and $C_A = C_V$, then $dV_A/dt = dV_V/dt$. This means the Miller factor should be "0" for same side switching.

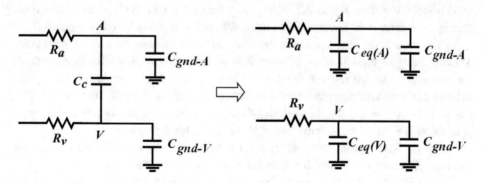

Fig. 2.6 Miller factor calculation

For the case where V_A and V_V having opposite transitions with $R_A = R_V$ and $C_A = C_V$, one obtains $dV_A/dt = -dV_V/dt$. Then a Miller factor of "2" is obtained for opposite switching. The Miller factor derived here is based on the worst-case assumption that signals at the victim and aggressor switch simultaneously with no skew.

Work in [4] considered differing slew rates for aggressor and victim line waveforms. Due to different driver strengths, we can have a fast rising/falling aggressor while having a slow victim waveform. This allows more charge to dump through the coupling capacitor to the victim. In this case, it is possible to obtain Miller factors as high as 3 for "opposite switching" and as low as −1 for "same direction switching" assuming saturated ramp voltages [4].

On the other hand, the coupling capacitor sees exponential waveforms rather than ramp waveforms on its terminals, the aggressor and victim waveforms should be represented by exponential functions rather than saturated ramp functions [8, 9]. Authors in [4] argued that Miller factors 3 and −1 corresponding to worst case and best case delay scenarios may not be correct upper and lower bounds for exponential type waveforms. The correct Miller factor values corresponding to exponential case need to be formulated. Therefore, next subsection derives Miller factors for worst-case and best-case scenarios for saturated exponential input waveforms and also formulates these factors for partial overlap cases.

2.2.1 Miller Factor Estimation Under Exponential Waveforms

The goal of this subchapter is to explore whether a Miller factor larger than 3 exists for "opposite switching" and also to find out if a lower bound than −1 exists for "same direction switching" under saturated exponential input waveforms.

Worst-Case Miller Factor Calculation
To examine worst-case scenarios, a piecewise linear approximation has been employed to represent rising and falling exponential waveforms due to the difficulties of working with exponential waveforms. Figure 2.7 shows how a rising unit exponential can be represented closely by a piece-wise linear approximation (dashed line) using only five segments.

In the model, the first segment represents the waveform in between $0 < t \leq 0.5t_r$. Similarly, 2nd segment is taken between $0.5t_r < t \leq t_r$ due to rapid changing nature of the waveform. The 3rd segment is taken between the t_r and $2t_r$ time points. Finally, the 4th and 5th segments approximate the waveform during $2t_r < t \leq 3t_r$ and $3t_r < t \leq 5t_r$ respectively. At $t = 5t_r$, it is assumed that the exponential waveform saturates. V_{DD} value is taken as 1 V for all waveforms during all Miller factor derivations.

Based on these values, the average error between the exponential function and the piecewise linear representation has been estimated to be less than 2.5%.

To obtain worst-case delay estimation, a strong aggressor is assumed implying a fast switching aggressor compared to victim driver as shown in Fig. 2.8. As shown, aggressor

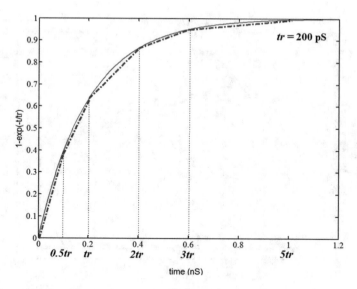

Fig. 2.7   Piece-wise linear representation for rising exponential waveform

waveform is taken as a rising exponential and the victim is assumed a falling exponential waveform. In order to have maximum delay effect on the victim, aggressor should complete its transition before victim reaches the 50% level, which is defined as victim threshold.

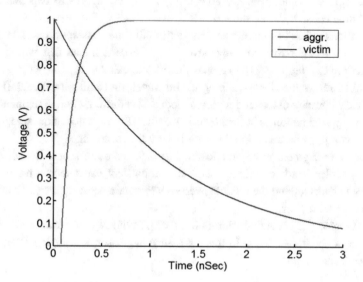

Fig. 2.8   Fast switching aggressor versus victim waveform

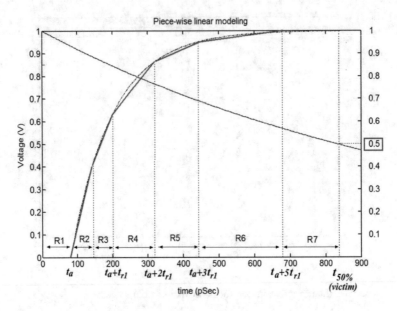

Fig. 2.9 Regions generated for Miller factor calculation

Figure 2.9 shows the close-up view of the aggressor waveform in Fig. 2.8, and indicates the six regions generated for Miller factor calculation. It is assumed that aggressor starts its transition t_a after victim waveform started its transitioning, and completes its switching before victim reaches the 50% threshold level. The rising aggressor exponential has an associated time constant t_{r1} as shown.

In derivations, it has been assumed that victim fall time constant t_{r2} is at least 7 times ($5t_{r1} = 0.69t_{r2}$) larger than the aggressor rise time constant so that whole aggressor transition completes before victim reaches the 50% threshold.

Figure 2.10 shows the slow switching victim waveform (falling exponential). Here, t_{r2} is the fall time constant associated with the victim waveform. Only one segment of piece-wise linear approximation is utilized during Miller Factor calculation since aggressor assumed to complete its transition during this very first segment.

Based on five-segment representation in Fig. 2.9, there are seven distinct regions to consider in Miller factor calculation. For this, equivalent capacitances for each region will first be calculated, and then a time average worst-case equivalent capacitance will be estimated over all regions.

The first region (R1) is defined during $0 < t < t_a$. In this region, victim starts falling but aggressor does not start until t_a. In this case since aggressor is at logic 0, the equivalent coupling capacitance is:

$$C_{eq-R1}(V) = C_C \tag{2.3}$$

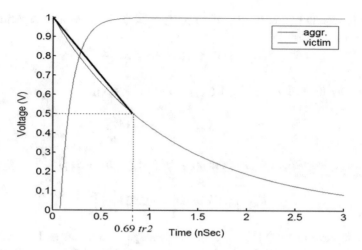

Fig. 2.10 One segment representation shown for the victim waveform

The second region (R2) is between $t_a < t \le (t_a + 0.5\,t_{r1})$, in this case we can refer to (2.2) for equivalent capacitance which can also be rewritten as:

$$C_{eq-R2}(V) = C_C\left[1 + \frac{|dV_A/dt|}{|dV_V/dt|}\right] \tag{2.4}$$

where,

dV_A/dt is the slope of corresponding aggressor segment,

dV_V/dt is the victim slope calculated from the first segment (Fig. 2.5).

Finally, for segment 1 we have:

$$C_{eq-R2}(V) = C_c\left[1 + \frac{0.393/0.5t_{r1}}{0.5/0.693t_{r2}}\right]$$

or

$$C_{eq-R2}(V) = C_c\left[1 + 1.09\frac{tr_2}{tr_1}\right] \tag{2.5}$$

The third region R3 (2nd segment) is defined during $(t_a + 0.5t_{r1}) < t \le (t_a + t_{r1})$, then:

$$C_{eq-R3}(V) = C_c\left[1 + \frac{(0.632 - 0.393)/0.5tr_1}{0.5/0.693t_{r2}}\right]$$

or,

$$C_{eq-R3}(V) = C_c\left[1 + 0.661\frac{tr_2}{tr_1}\right] \tag{2.6}$$

During the fourth region $(t_a + t_{r1}) < t \le (t_a + 2t_{r1})$, the equivalent capacitance:

$$C_{eq-R4}(V) = C_c \left[1 + 0.323 \frac{tr_2}{tr_1} \right] \tag{2.7}$$

Similarly for the fifth region $(t_a + 2t_{r1}) < t \le (t_a + 3t_{r1})$,

$$C_{eq-R5}(V) = C_c \left[1 + 0.119 \frac{tr_2}{tr_1} \right] \tag{2.8}$$

The sixth region (R6) is defined for $(t_a + 3t_{r1}) < t \le (t_a + 5t_{r1})$:

$$C_{eq-R6}(V) = C_c \left[1 + 0.035 \frac{tr_2}{tr_1} \right] \tag{2.9}$$

Finally, in seventh region (R7), aggressor voltage is constant at 1 V while victim decreases towards its 50% threshold point. Since, voltage difference across coupling capacitor C_c is less than the one in Region 1; we may expect an equivalent capacitance which is:

$$C_{eq-R7}(V) \le C_C \tag{2.10}$$

Since an average Miller factor over complete regions will be calculated, having this region will reduce the Miller Factor. However, the average worst-case Miller factor will be maximum if one eliminates R7 by increasing aggressor arrival time t_a. In this case, aggressor finishes its transition right before victim reaches its 50% voltage threshold. Based on this assumption, time average of worst-case equivalent coupling capacitance at victim node can be calculated:

$$
\begin{aligned}
&C_{eq}(V) \\
&= \frac{t_A(C_{eqR1}) + 0.5t_{r1}(C_{eqR2}) + 0.5t_{r1}(C_{eqR3}) + t_{r1}(C_{eqR4}) + t_{r1}(C_{eqR5}) + 2t_{r1}(C_{eqR6})}{0.693t_{r2}}
\end{aligned}
\tag{2.11}
$$

where,

$$t_A = 0.693t_{r2} - 5t_{r1}$$

After inserting (2.3), (2.5)–(2.9) into (2.11), one obtains the average worst case Miller capacitance at victim node:

$$C_{eq}(V) = 2.998 \cong 3.0C_c \tag{2.12}$$

which suggests that the worst case Miller factor is 3. This value is same as the one obtained for saturated ramp inputs in [4].

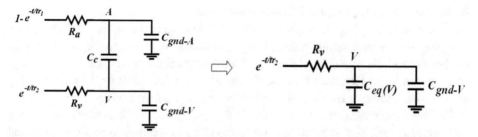

Fig. 2.11 Miller factor calculation for worst-case

This work has obtained the Miller factor for victim using a fast rising aggressor and slow falling victim waveform. Similar results can also be obtained for the case where a fast falling exponential aggressor and slow rising victim.

The above result can be further verified using a simple formulation that has been described below. Referring to Fig. 2.11, the current leaving the victim node via coupling capacitor C_c should be equal to the current on the equivalent victim capacitance $C_{eq}(V)$ for matching purposes, i.e.:

$$C_c\left(\frac{dV_V}{dt} - \frac{dV_A}{dt}\right) = C_{eq}(V)\frac{dV_V}{dt} \tag{2.13}$$

For obtaining worst case Miller factor, aggressor waveform can be assumed complete its transition before victim reaches to its threshold, which is taken typically as the 50% voltage value. The time that corresponds to the 50% point on victim waveform is $0.69t_{r2}$.

Assuming zero initial condition, both sides of above equation can be integrated over $0 \le t \le 0.69t_{r2}$ time interval. In other words, during this interval, we can equate the charges on both C_c and $C_{eq}(V)$:

$$C_c \int_0^{0.69tr_2} \left(\frac{dV_V}{dt} - \frac{dV_A}{dt}\right) = C_{eq}(V) \int_0^{0.69tr_2} \frac{dV_V}{dt} \tag{2.14}$$

Then, we have:

$$C_c[V_V(0.69t_{r2}) - V_V(0) - V_A(0.69t_{r2}) + V_A(0)] = C_{eq}(V)[V_V(0.69t_{r2}) - V_V(0)]$$

$$C_c(0.5 - 1 - 1 + 0) = C_{eq}(V)(0.5 - 1)$$

This finally gives $C_{eq}(V) = 3C_C$. This confirms the result that was previously obtained using (2.12).

Partial Overlap of Aggressor and Victim Waveforms

The worst case coupling delay effect occurs if a fast aggressor starts its rising transition after slow victim started transitioning, and completes its switching before victim reaches the 50% threshold level. This is assumed as the full overlap case. In reality, signals can start and end anytime. If the whole aggressor transition does not take place during $0 \le t \le 0.69 t_{r2}$, then we have less than a full overlap or a partial overlap.

The overlap duration can be expressed as $K_A * t_{r1}$, where K_A is overlap constant and t_{r1} is aggressor rise time constant. For full overlap, K_A is assumed to be 5 and for partial overlapping, it is in the range $0 < K_A < 5$.

In order to obtain a general formula for partial overlapping, we take piecewise segments at regular distances in time as shown in Fig. 2.7. Initially, $0.5 * t_{r1}$ distance has been selected for each segment. Figure 2.12 illustrates such a model with 10 different segments. The slow victim waveform is modeled using only one segment representation as in Fig. 2.10.

For partial overlapping scenario, K_A is in the range $0 < K_A < 5$, but has to be stated in multiples of 0.5 due to the resolution chosen in piece-wise linear model. Let us define the resolution by a constant r, which is equal to 0.5 in our case. Based on Fig. 2.12, we can then calculate an equivalent capacitance for each region (in case it exists):

For Region 1 during $0 < t < t_a$, we have:

$$C_{eq-R1}(V) = C_C \tag{2.15}$$

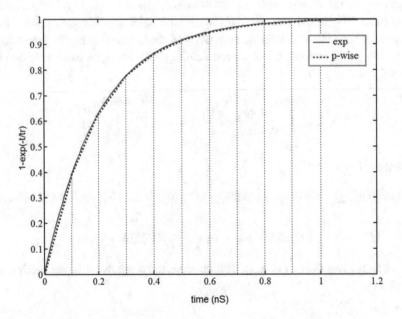

Fig. 2.12 Representation using 10-segments

During Region 2 which is between $t_a < t \le (t_a + 0.5\,t_{r2})$, we obtain:

$$C_{eq-R2}(V) = C_c \left[1 + \frac{(e^0 - e^{-0.5})/0.5t_{r1}}{0.5/0.693t_{r2}} \right] \qquad (2.16)$$

For Region 3, during $(t_a + 0.5t_{r1}) < t \le (t_a + t_{r1})$:

$$C_{eq-R3}(V) = C_c \left[1 + \frac{(e^{-0.5} - e^{-1})/0.5t_{r1}}{0.5/0.693t_{r2}} \right] \qquad (2.17)$$

Similarly, we can find the Miller factors for remaining eight regions. Knowing that above equations has a pattern, the time-averaged equivalent coupling capacitance for partial overlap case can be written:

$$C_{eq}(V)$$
$$= \frac{1}{0.69t_{r2}} \left[(0.69t_{r2} - K_A t_{r1})C_c + 0.5C_C \sum_{i=1}^{2K_A} \left(t_{r1} + 1.39 * \left(\frac{e^{\frac{-i+1}{2}} - e^{\frac{-i}{2}}}{0.5} \right) t_{r2} \right) \right] \qquad (2.18)$$

The formula above is valid for exponential waveforms at nodes A and V (Fig. 2.10) with victim fall time constant t_{r2} being at least seven times larger than the aggressor rise time constant t_{r1}. In this case, full aggressor transition takes before victim reaches the 50% threshold.

In most cases, overlap K_A may not be a multiple of 0.5. In this case, a smaller resolution constant r should be chosen (i.e. $r = 0.1$). The smaller value of r results in increased accuracy due increased resolution. In this case,

$$C_{eq}(V)$$
$$= \frac{1}{0.69t_{r2}} \left[(0.69t_{r2} - K_A t_{r1})C_c + rC_C \sum_{i=1}^{K_A/r} \left(t_{r1} + 1.39 * \left(\frac{e^{(-i+1)r} - e^{-ir}}{r} \right) t_{r2} \right) \right] \qquad (2.19)$$

where,

$0 < K_A < 5$ and $K_A = cr$ with $c = 1 \ldots 0.5/r$.

If overlap amount $K_A = 0$, a static capacitance of C_c is found as expected. With this formula, one can also verify the full overlap case which should give the worst-case Miller Factor. Assuming a resolution constant $r = 0.1$, for full overlap case with $K_A = 5$, one can obtain an equivalent coupling capacitance of $3C_c$ at victim node.

Best-Case Miller Factor Calculation

In this case, aggressor driver switches in the same direction as the victim driver, but transitions lot faster. Aggressor and victim voltages have been taken as falling exponential waveforms with fall time constants t_{r1} and t_{r2}, respectively. It is also assumed that aggressor driver is strong and it completes its transition before victim waveform reaches its 50% switching threshold (V_{ref}) ($t_{r1} < t_{r2}/7$).

According to Fig. 2.13, there are 6 distinct regions to consider during equivalent capacitance calculation. During Miller Factor calculation, only one segment of piece-wise linear approximation is considered for the slow switching victim waveform since aggressor assumed to complete its transition during this first segment for best-case delay (Fig. 2.10).

It is also assumed that victim and aggressor drivers start its falling transition simultaneously with $t_a = 0$. In this case, the maximum victim signal speed-up occurs [4]. Referring to Fig. 2.13, the first segment approximates the waveform between ($0 < t \leq 0.5 t_{r1}$). Then referring to (2.2), the equivalent capacitance can be rewritten as:

$$C_{eq-R1}(V) = C_C \left[1 - \frac{|dV_A/dt|}{|dV_V/dt|} \right] \quad (2.20)$$

The equivalent capacitance during first region (R1) can be written as:

$$C_{eq-R1}(V) = C_c \left[1 - 1.09 \frac{t_{r2}}{t_{r1}} \right] \quad (2.21)$$

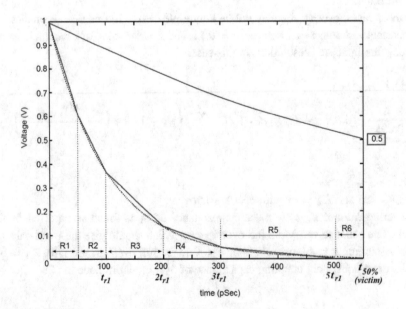

Fig. 2.13 Fast switching aggressor versus slow victim waveform

The second region (R2) is defined during $(0.5t_{r1} < t < t_{r1})$. For region 2:

$$C_{eq-R2}(V) = C_c\left[1 - 0.661\frac{tr_2}{tr_1}\right] \tag{2.22}$$

The remaining regions are defined as follows: The third region is taken between the t_{r1} and $2t_{r1}$ time points. During the fourth (R4) and fifth regions (R5), segments approximate the waveform during $2t_{r1} < t \leq 3t_{r1}$ and $3t_{r1} < t \leq 5t_{r1}$, respectively. At $t = 5t_{r1}$, it is assumed that the exponential waveform reaches to zero. The equivalent capacitances for 3rd, 4th and 5th regions can be calculated as follows:

$$C_{eq-R3}(V) = C_c\left[1 - 0.323\frac{tr_2}{tr_1}\right] \tag{2.23}$$

$$C_{eq-R4}(V) = C_c\left[1 - 0.119\frac{tr_2}{tr_1}\right] \tag{2.24}$$

$$C_{eq-R5}(V) = C_c\left[1 - 0.035\frac{tr_2}{tr_1}\right] \tag{2.25}$$

Finally, in the sixth region (R6), aggressor voltage remains constant at 0 V while victim waveform falls towards the 50% threshold point. We may expect an equivalent capacitance near C_c as the voltage difference on coupling capacitor is smaller. Based on this assumption, time average of best-case equivalent coupling capacitance at victim node can finally be calculated:

$$C_{eq}(V)$$
$$= \frac{0.5t_{r1}(C_{eqR1}) + 0.5t_{r1}(C_{eqR2}) + t_{r1}(C_{eqR3}) + t_{r1}(C_{eqR4}) + 2t_{r1}(C_{eqR5}) + (0.693t_{r2} - 5t_{r1})(C_{eqR6})}{0.693t_{r2}} \tag{2.26}$$

After substituting (2.21)–(2.25) into (2.26), the equivalent coupling capacitance at victim node for best-case delay found to be:

$$C_{eq}(V) = -C_c \tag{2.27}$$

This suggests that the best-case Miller factor is -1. These results suggest that the lower and upper bounds on Miller Factors for realistic exponential inputs remains same at -1 and 3 corresponding to best and worst case delay calculations.

We can further verify (2.27) by using a current formulation. Referring to Fig. 2.14, the current through coupling capacitor C_c should be same as current thru the equivalent victim capacitance $C_{eq}(V)$, namely:

$$C_c\left(\frac{dV_V}{dt} - \frac{dV_A}{dt}\right) = C_{eq}(B)\frac{dV_V}{dt} \tag{2.28}$$

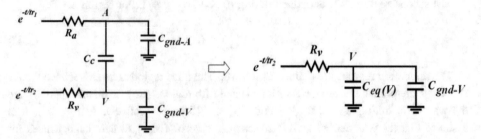

Fig. 2.14 Miller factor calculation for best-case

Assuming zero initial condition, we can integrate both sides of (2.28) over $0 \le t \le 0.69t_{r2}$ time interval:

$$C_c[V_V(0.69t_{r2}) - V_V(0) - V_A(0.69t_{r2}) + V_A(0)] = C_{eq}(V)[V_V(0.69t_{r2}) - V_V(0)]$$

$$C_c(0.5 - 1 - 0 + 1) = C_{eq}(B)(0.5 - 1)$$

This would again give $C_{eq}(V) = -C_C$ as in (2.27).

2.2.2 The Simulation Results

The Miller factors have been verified using industry standard HSpice software with parameter values derived in TSMC 130 nm technology. Two parallel wires on Metal3 have been assumed and were modeled by using many segments of π networks, i.e. every 100 μm wire is modeled using a π-network in HSpice. For a 1 mm wire, there were 10-π segments representing the RC distributed interconnect. Coupling capacitors were also distributed along the line as shown in Fig. 2.15.

The sheet resistance of metal interconnect in this technology is 74 mΩ/square. For a wire width is 0.2 μm, the per unit length capacitance and resistance values for each wire are 0.0226 fF/μm and 370 mΩ/μm, respectively. It has been assumed that aggressor and victim lines are driven by identical inverters of size 31.2 μm/10.4 μm and the loads at the end of the wires are also identically sized inverters.

The 50% victim delay has been first simulated using the two-line circuit with distributed coupling capacitances taken in between. Then, the coupled network has been reduced to an equivalent victim line with coupling capacitance scaled by a Miller factor (MF) as shown in Fig. 2.16. The 50% victim delay simulation has been repeated with the isolated victim line and compared results to coupled victim line delay to see how closely the MFs match to the coupled model.

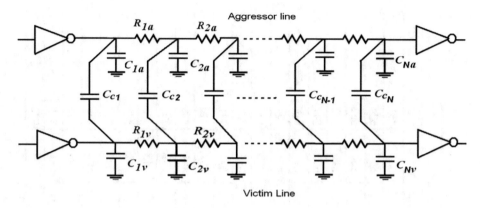

Fig. 2.15 Two parallel coupled interconnect network

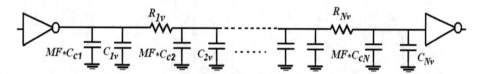

Fig. 2.16 The isolated victim line for delay calculation

In the simulation experiments, aggressor and victim interconnect resistances, ground and coupling capacitances have been varied in order to study the effects of varying interconnect lengths, widths and spacing. The parameter ranges are taken as follows: The aggressor and victim wire resistances are varied in between 20 and 250 Ω, the ground and coupling capacitances are changed from 50 to 300 fF. In addition, differing exponential rise/fall time constants on the inputs of both lines have been examined.

Table 2.1 summarizes the results obtained for worst case MF. For each case, aggressor arrival time has been adjusted for worst possible delay on the victim. For the 20 cases considered, the worst case MF of 3 produced very accurate delay results (an average error of only 2.69%) compared to coupled RC network delay.

Almost for all cases, there was no underestimation on worst-case delay when using a MF of 3 for exponential inputs. The results for the MF of 4 have also been included as this was suggested earlier in [4]. Results show that a factor greater than 3 results in overly pessimistic worst-case delay calculations for coupled RC interconnects and should be avoided. For some cases, an equivalent coupling capacitor of $4C_c$ gives an error as large as 14% compared to the coupled RC network.

Table 2.2 shows the 15 different cases examined for Best-Case MF. Previously, for ramp inputs a MF of -1 was suggested as a bound for best-delay calculation [4, 5]. However, the verification on the negative MF were never been done. Here, the results for

Table 2.1 Experimental results obtained for the worst case Miller factor

Case #	Agg/victim rise/fall time cons.	R_{agg} (Ω)	R_{vic} (Ω)	C_{agg} (fF)	C_{vic} (fF)	C_c (fF)	Delay in ps		Coupled RC network
							MF = 3	MF = 4	
1	30/300	100	120	50	60	100	352	382	351
2	80/600	100	100	60	60	140	627	669	624
3	50/450	160	80	90	60	160	522	564	522
4	50/400	80	70	60	80	150	477	515	472
5	60/500	250	160	120	100	130	560	603	547
6	60/500	160	240	100	120	200	638	693	616
7	100/750	30	20	100	120	170	777	824	781
8	100/750	70	60	140	120	220	827	873	821
9	100/750	150	150	100	100	150	773	839	768
10	60/700	60	80	150	200	170	771	817	771
11	60/700	100	140	250	300	140	784	832	779
12	100/800	20	25	100	100	200	837	891	839
13	50/600	50	50	60	60	280	733	789	725
14	40/400	100	120	100	150	300	609	642	585
15	75/600	200	200	100	100	50	557	577	553
16	100/800	100	100	100	100	200	851	909	843
17	50/400	220	200	80	80	120	468	503	455
18	40/550	20	20	70	80	180	620	663	615
19	50/400	100	120	100	150	250	575	631	555
20	100/800	100	200	80	220	300	994	1070	972

victim speed-up under exponential inputs for the coupled RC circuit will be shown as well as for the isolated victim network. As usual, the 50% voltage point has been taken as reference in delay calculation.

In simulation experiments, aggressor and victim interconnect resistances, ground and coupling capacitances have been varied in order to study the effects of varying interconnect lengths, widths and spacing and examine differing exponential time constants on the inputs of both lines. The results show that for the cases considered absolute error remains less than 7% for all cases, and the average error obtained is only 3.3%.

2.2.3   Summary

The change in wire delay due to cross-coupling noise can have deleterious effects on performance in nanoscale technologies. The increased delays occur when an aggressor line switches in the reverse direction to the victim line. An aggressor can also cause decreased delays when switching in the same direction as the victim. The increase/decrease in delay can violate the setup or hold time requirements for logic storage circuits.

The crosstalk induced delay estimation using traditional simulation tools such as Industry standard HSpice is computationally expensive and inapplicable to full-chip analysis. Alternative techniques are desirable to ensure signal integrity in a limited design cycle time. The Miller factors can be used to approximate a coupling capacitance by two grounded capacitances so that the conventional delay estimation methods can still be utilized in a limited design cycle.

2.3   Interconnect Models

The wire parameters can have its impact on the electrical behavior and affects circuit delay, power dissipation and reliability. In order to study these effects, electrical models are used to approximate the real behavior of a wire as a function of its parameters.

The Lumped Model
Normally, the wire parasitic is distributed along its length and is not lumped to a single position. Ideally, a distributed model should be used to represent interconnect. However, due to the complexity issues, the behavior is mostly approximated by sections of lumped circuit elements.

However, if only one single parasitic effect is dominant or we need a very quick estimate, we can lump different fractions into a single element. In this case a single differential equation can describe the parasitic effects.

In the "*Lumped*" modeling, the distributed capacitance is lumped into one single capacitance while ignoring the line resistance. Figure 2.17 shows the *Lumped C* representation.

Table 2.2 Experimental results obtained for the best case Miller factor

Case #	Agg/victim rise time constant	R_{agg} (Ω)	R_{vic} (Ω)	C_{agg} (fF)	C_{vic} (fF)	C_c (fF)	Delay with MF = −1 (ps)	Coupled RC (ps)	|error %|
1	30/300	100	120	50	60	100	243	256	5.08
2	80/600	100	100	60	60	100	452	463	2.38
3	50/450	160	80	90	120	100	337	343	1.75
4	50/400	80	90	160	180	250	316	340	7.06
5	60/500	250	160	120	100	130	391	416	6.01
6	60/500	160	240	100	140	75	443	450	1.56
7	100/750	30	20	100	120	170	550	555	0.90
8	100/750	70	80	140	170	220	544	570	4.56
9	100/750	150	150	100	100	150	557	573	2.79
10	60/700	60	80	150	200	170	557	563	1.07
11	60/700	100	140	250	300	110	633	640	1.09
12	100/800	20	25	100	100	150	505	518	2.51
13	100/800	100	100	100	100	150	590	595	0.84
14	100/800	100	200	110	240	300	576	616	6.49
15	120/900	200	200	100	100	250	622	655	5.04

Fig. 2.17 Lumped C modeling

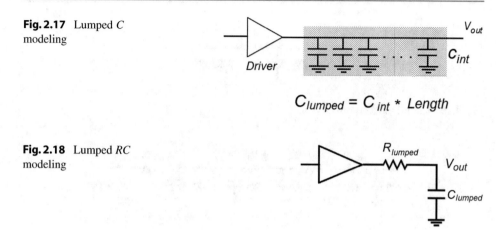

$$C_{lumped} = C_{int} * Length$$

Fig. 2.18 Lumped RC modeling

As indicated in the figure, the lumped capacitance value is calculated after multiplying the per unit length capacitance C_{int} by the length of the wire. The only impact of the wire would be the loading effect on the driver gate.

This model is used for a rapid estimation of wire delay, although it is quite inaccurate. Lines themselves possess resistance and resistance impact cannot be ignored.

An improved version would be the *Lumped RC* model (see Fig. 2.18). In this model, the distributed resistance is lumped into one single resistance R and similarly the different fractions of capacitance are lumped into one single capacitance C.

For a step function applied at the driver output (having a magnitude of V_{DD}), one can expect a rising exponential on the output:

$$V_{out}(t) = V_{DD}\left(1 - e^{-t/(R_{lumped}*C_{lumped})}\right) \qquad (2.29)$$

from which the propagation delay can easily be calculated by setting left hand side equal to $0.5 * V_{DD}$:

$$\tau_{PLH} = 0.69 R_{lumped} C_{lumped} \qquad (2.30)$$

Although this model is often used as a first order calculation to get an upper bound on delay, the delay calculation using *lumped RC* model overestimates the actual delay hence paints a pessimistic picture on circuit performance.

The π-Model

The simple lumped RC model provides a very rough approximation of the actual transient behavior of interconnect. The accuracy of the simple lumped RC model can be improved by dividing the total line capacitance into two equal parts, as shown in Fig. 2.19. This representation is named as the π-model representation [10].

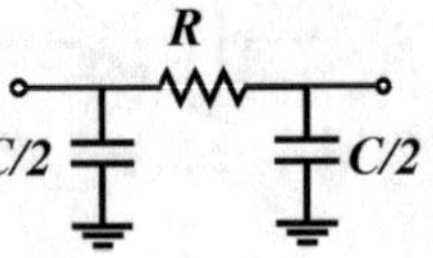

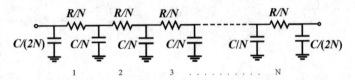

Fig. 2.19 The π-model representation

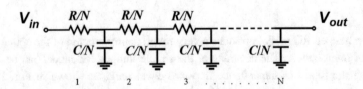

Fig. 2.20 N-segment π representation

More accurate representations of the distributed RC circuit can be obtained by using multiple-π segments. In an N-π representation, the total resistance and capacitance values of each π-segment are given as R/N and C/N, respectively. Since the capacitors are split equally on each side, the ground capacitors are calculated as $C/2N$ for each π-segment. The capacitor values at intermediate nodes would be doubled due to the neighboring π segment (Fig. 2.20).

The accuracy of this model increases with increasing N, where the transient behavior approaches that of a distributed *RC* line for very large values of N. Kawaguchi and Sakurai have reported that the error in simulating delay of the distributed RC line by using 5-π segment is less than 1% for almost all cases [11].

The RC ladder network representation is also preferred in circuit simulations. Similar to the N-π modeling, the accuracy of the model increases as N goes very large. Figure 2.21 shows the RC ladder model. This model is based on the simple lumped RC model.

In calculating the delay for representations other than simple lumped *RC* model, especially in tree networks, Elmore Delay formulation comes very handy as it provides a very powerful and quick estimation tool. RC trees are commonly used to model digital logic gates and their associated interconnect paths at various stages.

Fig. 2.21 N-segment RC ladder representation

2.4 Elmore Delay Formulation

Elmore delay provides a simple approximation of delay [12] and provides itself as a useful tool in timing optimization. It provides a simple closed-form expression and provides an upper bound on circuit delay.

The actual delay of a line asymptotically approaches Elmore delay as input signal rise time increases. This technique can be applied to tree-like circuit structures where there are no loops. Since most circuits can be represented as an RC circuit with no loops, the technique becomes very useful in delay estimation.

Elmore delay estimates the delay from a source (root) to one of the leaf nodes. It assumes that the network has a single input and all capacitances should also appear between a particular node and the ground. In addition, there should not be any resistance loops. In other words, the circuit should have a tree-like structure similar to the one shown in Fig. 2.22.

In order to explain Elmore delay, we need to have some definitions. The first definition is the *path resistance*. Each node here has a path resistance R_{nn} which equals to the sum of resistance coming from the input node.

As an example, for node 4, the path resistance equals $R_1 + R_4$. Similarly, for the output node the path resistance would simply be $R_1 + R_2 + R_5 + R_6$.

The common or *shared path resistance*, on the other hand, is defined between two nodes. As the name implies, the shared path resistance between node n and node m, is the simple intersection of path resistances R_{nn} and R_{mm}.

When calculating the delay between the input and any node, the equation simply multiplies each ground capacitance with the shared resistances. Elmore delay at node n is given by:

$$\tau_{delay-n} = \sum_{k=1}^{N} C_k R_{nk} \tag{2.31}$$

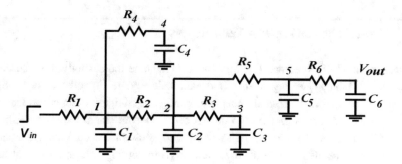

Fig. 2.22 Elmore delay example

In this formulation, it has been assumed that each node is initially at zero, and a step input is applied at t = 0. In the formula, N indicates the number of nodes in the circuit. This means each capacitance gets multiplied by the shared path resistance between that node and the desired node.

Referring to Fig. 2.22, assume we want to calculate delay between input node and the leaf node 3. In this case, we multiply each capacitance by the shared resistance between that node and node 3. Starting with node 4, capacitance C_4 hence gets multiplied by R_1 only. This is because while its path resistance is $R_1 + R_4$, the path resistance to node 3 is $R_1 + R_2 + R_3$ and hence the intersection gives only one resistance which is R_1.

As for C_5 (node 5), the shared resistance is $R_1 + R_2$, hence it produces the product term $C_5(R_1 + R_2)$. The delay formula for node 3 is given as:

$$\tau_{delay-3} = C_1 R_1 + C_2(R_1 + R_2) + C_3(R_1 + R_2 + R_3)$$
$$+ C_4 * R_1 + C_5(R_1 + R_2) + C_6(R_1 + R_2)$$

The delay on the output node V_{out} can similarly be formulated using the same principle:

$$\tau_{delay-V_{out}} = C_1 R_1 + C_2(R_1 + R_2) + C_3(R_1 + R_2) + C_4 * R_1$$
$$+ C_5(R_1 + R_2 + R_5) + C_6(R_1 + R_2 + R_5 + R_6)$$

Note that the Elmore delay or τ_{delay} defines the time point where the output waveform its 63% of the maximum value. This is different than the propagation delay where it is defined at 50%.

Elmore delay can be used in deriving the distributed wire delay. For this, one can use the RC ladder network and then take $N \to \infty$. After some steps, it can easily be shown that the propagation delay of a distributed (actual) wire in fact approximated at:

$$\tau_{PLH-actual} = 0.38 R_{lumped} C_{lumped} \qquad (2.32)$$

2.5   Crosstalk Mitigation Techniques

As stated in the ITRS Roadmap [1], interconnects become increasingly more susceptible to noise as technology scaling further into nanoscale regime. In order to deal with the challenges associated with crosstalk noise estimation and mitigation, tools and techniques should be included early into the chip design cycle. During signal integrity verification, it is essential that crosstalk noise pulse width and peak values on sensitive nodes are below the recommended threshold levels and a given routing solution will not lead to logic failures caused by the coupled noise.

2.5.1 Driver Sizing

Victim Driver Sizing method is a practical crosstalk noise reduction technique in the post route design stage [2, 13, 14]. If a victim driver is sized-up, the drive strengths of its NMOS and PMOS transistors increase and it can more effectively hold the victim net at a steady voltage and the increased device capacitance also adding to the stability of the victim line. As a result, noise induced by the affecting wire (aggressor) on a victim net would be less. The victim driver sizing is also a very effective means to reduce crosstalk delay.

In *Aggressor Driver Sizing* methodology, the aggressor size is reduced down to mitigate the crosstalk. In this case, its effective transconductance decreases, and the aggressor driver cannot achieve transition as fast due to its large resistance. As a result, noise amount induced on victim line would be less.

The difficulty in gate sizing for noise reduction is that, by changing the driver size, noise at the driver output is reduced, but noise injected by that driver on other nets is increased. In the case of victim driver sizing, one might want to also consider the situation where a victim net can also become an aggressor from the opposite standpoint. The driver sizing can create cyclical dependencies between nets in the circuit with noise violations [13].

Hashimoto and Onodera [15] suggested a driver-sizing algorithm where the goal is to minimize the sum of squared noise voltages at the aggressor and victim. However, when this optimization was applied to victim driver sizing method, it may result in larger driver sizes than needed. This also means larger power dissipation and increased IC temperature.

The driver sizing methodology requires moderate to large extra circuit area and affects the timing of paths through the gate. Hence, designers need to be careful when applying this technique.

2.5.2 Wire Spacing

Wire spacing is another effective technique to mitigate crosstalk. For a fixed wire width, if the wire spacing to its neighbors is increased, its coupling capacitance would be decreased. Considering a parallel plate relation between the wire and its neighbor, there is an inverse relation between the distance d and the capacitance C, which is expressed by $C = \varepsilon_0 A/d$ where A is the area of the plate and ε_0 is the permittivity. The more the spacing would be between the lines, the less crosstalk noise would be produced on the victim line.

Although wire spacing is an effective technique, it results in large area penalty due to the use of routing resources.

2.5.3   Wire Sizing

As a wire's width is changed, its resistance and capacitance values also vary. Larger wire sizes or widths mean reduced wire resistances and increased ground capacitances, all of which contribute to victim stability [16]. For example, if the victim driver is holding the victim line at steady (i.e. GND), it will be more effective holding it because of reduced line resistance and increased capacitances, and hence the crosstalk noise induced on the victim will be less. Wire sizing reduces the noise amplitude although not as strong as wire spacing.

Wire sizing, on the other hand, may not reduce crosstalk delay since sizing increases the ground capacitance of the victim line. In some cases, wire sizing may result in an increase in victim line delay.

2.5.4   Power Supply Shielding

In *Shielding* method, a voltage supply or ground interconnect can be inserted between the aggressor and victim interconnects to reduce the impact of crosstalk (see Fig. 2.23). Since V_{DD} and ground interconnects are routed normally for powering the logic and test circuits; it may be manageable to reroute these power interconnects to provide shielding for interconnect affected by the crosstalk noise. This approach does have the drawback of increasing the victim total ground capacitance and can have increased delay effects on victim.

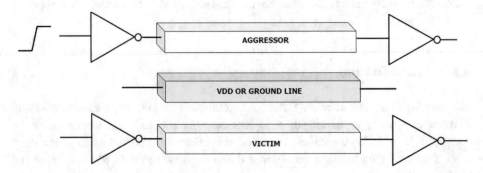

Fig. 2.23 Shielding method

2.5.5   Buffer Insertion Method

The coupling capacitance amount is dependent on distance the two lines can run in parallel to each other. By inserting a buffer, however, the coupling capacitance can be distributed into two parts due to two newly created wires. The resulting noise pulse would be smaller and hence can be mitigated easily due to filtering ability of receiver gates. Figure 2.24a shows a case where the crosstalk noise is greater than the tolerable noise margin of the sink. As a result, a fault appears on the output. On Fig. 2.24b, a buffer is inserted at the mid-point on the coupling portion. Due to this, the amplitude of the noise pulse is now lower the sink's noise margin, and the circuit functions correctly.

Inserting a buffer can sometimes be difficult since it may introduce additional delay and can result in a timing violation in the critical path. Buffer insertion also increases power consumption.

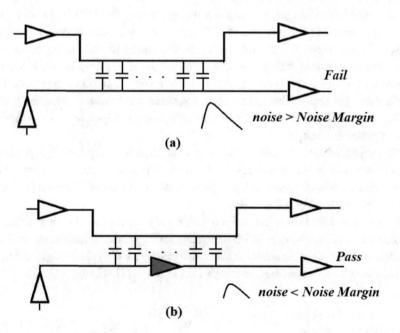

Fig. 2.24   Buffer insertion

2.5.6 A New Methodology for Crosstalk Reduction Using Transmission Gates

This crosstalk noise mitigation methodology uses on a novel transmission gate (TG) design with minimum area and performance overheads [17].

Noise filtering techniques based on TGs have normally been proposed for mitigating Single Event Transients or SETs. Similar to a crosstalk pulse, an SET pulse can pass through a series of gates and reach storage elements under certain conditions. If the generated pulse arrives at the storage element during its latching window, incorrect data can be stored resulting in soft error [18].

Generally speaking, SET filtering methods such as driver sizing (on-site filtering) and low pass filtering techniques (along the data path) [18, 19] could also be exploited for interconnect crosstalk mitigation. Here, the applicability of the TG based techniques will be explored for crosstalk noise mitigation.

In the basic TG technique, the pass transistors forming the TG are always made conducting by connecting the gates of PMOS and NMOS pass transistors to V_{SS} (ground) and V_{DD} (power supply), respectively [20]. In this method, while data signals can pass through the circuit with little change, the magnitude of transient pulses is reduced by the circuits. Technique becomes useful after using a two stage TG configuration with the condition that the size of the second stage transistor widths are 4 times larger than the first TG stage as shown in Fig. 2.25. In this case, a 40% reduction on the magnitude of transient pulses is reached.

This method has the limitation that it cannot eliminate large voltage glitches [17]. Another limitation is that it imposes a 5X area overhead regardless of the magnitude of crosstalk that need to be eliminated. Hence, the area overhead can be expensive especially when eliminating weaker noise pulses.

The new crosstalk mitigation method, however, is based on applying different gate biases on the pass transistors of a TG depending on the crosstalk magnitude that needs to be mitigated. Since filtering ability is controlled simply by the gate voltage, this approach requires a smaller area overhead in comparison to basic TG scheme.

Fig. 2.25 Transmission gates acting as low pass filters in mitigating noise pulses

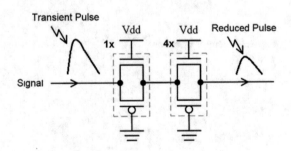

Normally to operate a TG, an NMOS gate voltage (V_{GN}) of V_{DD} and a PMOS gate voltage (V_{GP}) of 0 V is used. A TG can be still activated with a V_{GN} less than V_{DD} and a V_{GP} greater than 0 V as long as the gate to source-voltage (source-gate) voltage is greater than V_{TN} ($|V_{TP}|$), where V_{TN} and V_{TP} are NMOS and PMOS transistor threshold voltages, respectively [17].

The simulation results show that the scheme with varied gate voltage offers improved filtering characteristics in comparison to the basic TG method. The delay penalty of new TG method is proportional to crosstalk pulse-width that needs to be eliminated. If this is not acceptable, the technique can also be combined with the gate sizing method in mitigating the crosstalk pulse to keep delay overhead under control. In this case, driver sizing is used only partially to suppress the propagated noise at the output of the victim driver while the TG removes the remaining crosstalk glitch. The delay of the TG can be controlled by applying different gate biases of transistors in the TG.

Transmission Gate Filter Design

The new method is based on applying different gate biases on the TG pass transistors depending on the crosstalk noise magnitude that needs to be mitigated. The TG is placed at the end of the victim line before victim driver.

In comparing this technique with the basic TG technique [20], the setup shown in Fig. 2.26 has been utilized. The crosstalk noise in this figure is modeled using a trapezoidal waveform with rise and fall times of 15 ps at the input of the TG. This approximation is done for simulation convenience as the proposed technique is compared with the basic TG scheme. In simulations, the width of the trapezoidal waveform is measured about $0.5V_{DD}$ where V_{DD} is the power supply voltage. This circuit has been designed in 45 nm process technology using Predictive Technology Model (PTM) [21] and simulated using HSpice. In Fig. 2.27, V_a and V_b denote the output voltages of the TG stage and first inverter, respectively.

A 100 ps crosstalk pulse duration has been selected at the input and applied using the V_{in} voltage source. For the basic TG method, the gate voltages are taken as V_{GN} = 1.2 V (V_{DD}), and $V_{GP} = 0$ V. For the new method, the gate voltages of NMOS and

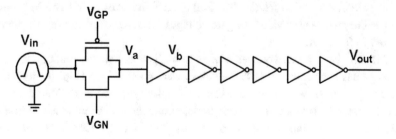

Fig. 2.26 The simulation set-up

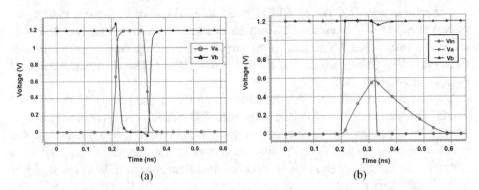

Fig. 2.27 Waveforms for **a** $V_{GN} = 1.2$ V and $V_{GP} = 0$ V and **b** $V_{GN} = 0.7$ V and $V_{GP} = 0.5$ V

PMOS transistors are varied symmetrically from their respective values using $V_{GN} = V_{DD} - \Delta V$ and $V_{GP} = \Delta V$, where $\Delta V > 0$. In simulations, ΔV value has been changed in increments of 100 mV until noise mitigation has taken place. Figure 2.27 compares the output waveforms for the basic TG gate scheme and the new methodology for chosen $\Delta V = 0.5$ V.

Referring to Fig. 2.27a, it is evident that the basic TG method cannot mitigate the transient. On the other hand, the TG with $V_{GN} = 0.7$ V and $V_{GP} = 0$ can mitigate the 100 ps crosstalk pulse (Fig. 2.27b). The voltage at V_b remain almost constant near V_{DD}.

The strong filtering effect can be attributed to increased resistance of the pass transistors on the TG due to reduced gate-source (source-gate) voltages of NMOS (PMOS) transistors. Figure 2.28 shows the TG resistance as a function of output voltage when gate voltages (V_{GN}, V_{GP}) are varied up to 0.2 V in steps of 0.1 V.

By simply adjusting the individual pass transistor's gate voltages, one can eliminate various crosstalk pulse durations. These gate voltages need to be generated internally using some extra hardware or provided off-chip using I/O pads. In generating the gate voltages, one idea is to use a simple voltage divider circuit consisting of serially connected devices [22]. Figure 2.29 shows a configuration where a supply voltage-divider with diode-connected NMOS transistors operating in saturation region can be used to obtain the gate-bias voltages for the TG. In this case, the sizes of M1 and M2 transistors are designed to give desired values of gate voltages. The area overhead for the bias circuit is expected to be small.

In Table 2.3, the TG gate voltages needed to mitigate certain crosstalk pulse durations up to 200 ps has been illustrated. These values were obtained via HSpice simulations. In simulations, both positive and negative crosstalk pulse types were considered.

For a TG filter designed, it is possible to determine the maximum noise width that can be mitigated, and the minimum width for signals that can be preserved. The difference in these two values can be defined as the TG filter margin [17]. Referring to Table 2.3, the

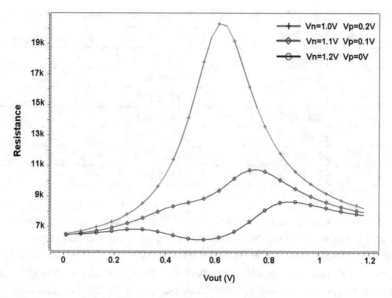

Fig. 2.28 The resistance of the TG as a function of output voltage

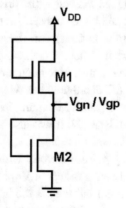

Fig. 2.29 The NMOS voltage divider circuit

filter margin varies between 12 up to 32% of the maximum noise pulse-width, indicating the clean cutoff properties of the proposed TG filter.

Transmission Gate Combined with Driver Sizing
The delay penalty of the proposed TG method is proportional to crosstalk pulse width that needs to be mitigated. For faster designs, however, the delay penalty may become unacceptable. In this case, the TG technique can be applied along with the driver sizing

Table 2.3 Gate voltage combinations needed for filtering various crosstalk noise widths

Crosstalk PW (ps)	Gate voltages V_{GN}/V_{GP} (V)	Signal duration (ps) positive/negative pulse
50	0.89/0.36	56/58
75	0.76/0.47	92/95
100	0.71/0.52	128/134
125	0.67/0.55	161/169
150	0.65/0.57	200/208
175	0.64/0.58	232/237
200	0.63/0.59	265/268

method to keep delay penalty manageable and the mitigation can be shared between the TG and the driver sizing. In this case, victim driver sizing method is used partially to suppress the crosstalk noise while the TG eliminates the remaining pulse. In order to keep delay penalty manageable, the TG delay can be controlled using different gate biases on the pass transistors. This would be determined depending on the available time-slack on a given logic path. In comparison to driver sizing taken alone, this combined approach can save considerable circuit area.

In order to compare the combined method with "driver sizing" methodology and the "basic transmission gate", two parallel interconnects on the very top layer in 45 nm technology have been considered with wire dimensions taken as follows [1]: the width (W), spacing (S) is 0.103 μm, and wire thickness T is 0.236 μm.

In simulations, a 10-π model with distributed coupling capacitances is used for every 0.5 mm of wire to represent the RC distributed behavior (see Fig. 2.30). The inductance effects have been ignored and the capacitive coupling has been assumed as the dominant mechanism for crosstalk in the analysis [2]. It was also assumed that the loads at the end of wires are identically sized inverters.

Referring to Fig. 2.30, a falling pulse waveform with 3 ps fall time has been applied to aggressor driver while victim driver was kept at V_{DD}. This would simulate the positive crosstalk pulse experiment. The small fall time at the input of the aggressor ensures that maximum amount of noise is induced on the victim net. For negative crosstalk glitch, the same setup is used except that a rising pulse waveform is tied to aggressor driver and a 0 V is applied to victim driver input.

In order to create various aggressor/driver scenarios, the aggressor driver size has been changed from 18X up to 100X (where X is the minimum inverter size) while victim driver sizes were varied from 2X up to 6X. The victim driver sizes were selected such that large crosstalk magnitudes occur at the victim line and propagate thorough victim receivers. Different wire lengths from 0.5 up to 3 mm were also examined.

In determining the delay and area overheads, the victim driver sizing methodology has been first considered. In order to mitigate the crosstalk noise, the victim driver size has

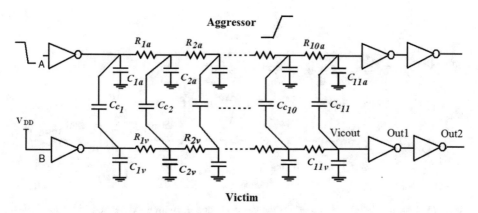

Fig. 2.30   The simulation setup used for crosstalk

been increased in steps of 0.5X until no crosstalk pulse appears at the receiver output. Once the driver size has been determined, the circuit in Fig. 2.30 has been reconstructed using the new victim driver size to perform a delay simulation. The simulation results show that victim driver sizing results in signal speedup and do not incur any delay overhead.

Figure 2.31 shows the circuit setup used for the basic TG and the combined method. The combined method relies on using only a single TG while the basic TG methodology uses a two stage TG configuration. For the basic TG technique, a single stage itself did not provide a meaningful reduction on the crosstalk pulse. Hence, a two-stage configuration has been considered during simulations. For this technique, the victim driver has been sized only if the 2-stage configuration proved to be unsuccessful in mitigating the crosstalk transient.

The proposed methodology is based on the combined use of driver sizing and the TG with varied gate biases. Normally, the TG design with varied gate bias is able to mitigate any crosstalk noise without the need for any driver sizing. The combined approach is only necessary for timing critical paths in order to keep delay penalty low.

In simulations, the available time-slack for victim critical path is assumed to be only 80 ps. The simulations have shown that 1X increase on victim driver size is sufficient to keep delay at manageable at this given slack. Hence, the combined method relies on incrementing the victim driver size by a certain amount (in our case by 1X) and then choosing the proper gate voltages of TG to eliminate the remaining transient. This allows

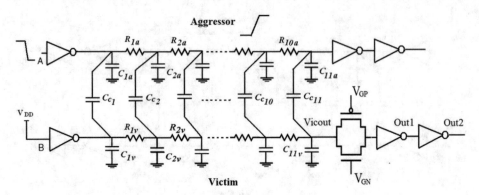

Fig. 2.31 The simulation setup used for the basic TG and the combined method

one to use a weaker low pass filter (a TG with V_{GP}/V_{GN} closer to $0/V_{DD}$, respectively) compared to the case that the TG is used alone.

Table 2.4 shows the comparison results for the first 20 cases considered. In this analysis, both positive and negative crosstalk pulses were considered. The analysis results show that for most cases, the basic TG method always results in more area overhead compared to driver sizing taken alone. The area overhead for the proposed method, on the other hand, remains constant at 2X assuming gate voltages are all provided off-chip via input pins. Compared to victim driver sizing alone, the proposed approach saves 47% in circuit area.

According to Table 2.4, the delay overhead for the proposed method remains below 73 ps for all cases which is below the available timing slack of 80 ps. If a lower delay overhead is needed, the victim size can be further increased in the combined method.

In this case, the TG gate voltages would be determined based on the glitch magnitude remaining after the sized up victim driver. Although increasing the victim driver size more than 1X (e.g. 1.5X) would reduce the area benefit to some extent, it would still save circuit designers considerable area.

A total of 500 cases were examined. According to the results, the combined approach results in on average a 41% area savings compared to victim driver sizing method taken alone. The delay overhead, on the other hand, remained less than 80 ps for all cases.

The combine TG and driver sizing approach results in considerably smaller area overhead and requires minimum change in circuit layout.

Table 2.4 The simulation results for the first 20 cases

Case #	Aggressor driver size	Victim driver size	Line length (mm)	Glitch type	Crosstalk magnitude (mV)	Area overhead (min inverter size)			Delay overhead (pS)		
						Driver sizing	Basic TG	Proposed method	Driver sizing	Basic TG	Proposed method
1	100X	2X	0.5	+	810	4X	5X	2X	−259	69	−66
2	100X	2X	0.5	−	810	3X	5X	2X	−235	71	−99
3	100X	3X	1.0	+	800	6X	6X	2X	−286	−3	73
4	100X	3X	1.0	−	790	3.5X	6.5X	2X	−223	−44	−41
5	80X	6X	1.5	+	730	4.5X	5X	2X	−137	105	44
6	80X	6X	1.5	−	730	3X	5X	2X	−112	64	−10
7	50X	2X	2.0	+	790	7.5X	8X	2X	−577	−315	−55
8	50X	2X	2.0	−	800	6.5X	8.5X	2X	−682	−290	−114
9	50X	3X	2.5	+	790	7.5X	7.5X	2X	−727	−336	−7
10	50X	3X	2.5	−	790	6X	7X	2X	−768	−507	−63
11	45X	4X	3.0	+	710	5.5X	7X	2X	−531	−175	42
12	45X	4X	3.0	−	690	4.5X	8X	2X	−538	−364	−49
13	30X	3X	2.0	+	730	4.5X	6.5X	2X	−495	−140	−42
14	30X	3X	2.0	−	780	3.5X	6.5X	2X	−541	−273	−133
15	30X	2X	2.0	+	800	5.5X	6.5X	2X	−617	−367	−91
16	30X	2X	2.0	−	800	4.5X	8X	2X	−689	−392	−152
17	20X	3X	0.5	+	700	2X	5X	2X	−60	−82	20
18	20X	3X	0.5	−	570	2.5X	5X	2X	−39	58	47
19	18X	2X	1.0	+	740	2.5X	5.5X	2X	−356	0	−109
20	18X	2X	1.0	−	760	2X	5X	2X	−376	84	−139

References

1. International Technology Roadmap for Semiconductors (ITRS), "Interconnect", 2011, http://www.itrs.net/.
2. S. Sayil, U. K. Borra, "A Multiline Model for Time-Efficient Estimation of Crosstalk," *Analog Integrated Circuits and Signal Processing*, vol. 59, no. 1, pp. 65–75, 2009.
3. S. S. Sapatnekar, "Capturing the effect of crosstalk on delay", *Proceedings of 13th Int. Conf. on VLSI Design*, pp. 364–369, 2000.
4. A. B. Kahng, S. Muddu, E. Sarto, "On switch factor based analysis of coupled RC interconnects", *Proc. of Design Automation Conf.*, pp. 79–84, 2000.
5. P. Chen, D.A. Kirkpatrick, K. Keutzer, "Miller factor for gate-level coupling delay calculation", *Int. Conf. on C. A. D. (ICCAD)*, pp. 68–74, 2000.
6. P. F. Tehrani, S. W. Chyou, and U. Ekambaram, "Deep submicron static timing analysis in presence of crosstalk," *Proc. ISQED*, pp. 505–512, 2000.
7. B. Franzini, C. Forzan, D. Pandini, P. Scandolara, and A. Dal Fabbro, "Crosstalk aware static timing analysis: a two step approach," *Proc. ISQED*, pp. 499–503, 2000.
8. T. Sato, Y. Cao, K. Agarwal, D. Sylvester, and C. Hu, "Bidirectional closed-form transformation between on-chip coupling noise waveforms and interconnect delay change curves," *IEEE Trans. on CAD*, vol. 22, no. 5, pp. 560–572, 2003.
9. K. Agarwal, Y. Cao, T. Sato, D. Sylvester, and C. Hu, "Efficient generation of delay change curves for noise-aware static timing analysis", in *Proc. of Asia and South Pac. Design Automation Conf.*, pp. 77–84, 2002.
10. M. R. Becer, D. Blaauw, V. Zolotov, R. Panda and I. N. Hajj, "Analysis of noise avoidance techniques in DSM interconnects using a complete crosstalk noise model," *Proceedings 2002 Design, Automation and Test in Europe Conference and Exhibition*, 2002, pp. 456–463.
11. H. Kawaguchi and T. Sakurai, "Delay and noise formulas for capacitively coupled distributed RC lines," *Proceedings of 1998 Asia and South Pacific Design Automation Conference*, 1998, pp. 35–43.
12. J. Rubinstein, P. Penfield, Jr., and M. A. Horowitz, "Signal delay in RC tree networks," *IEEE Transactions on Computer-Aided Design*, CAD-2(3), pp. 202–211, July 1983.
13. M. R. Becer, D. Blaauw, V. Zolotov, R. Panda and I. N. Hajj, "Analysis of noise avoidance techniques in DSM interconnects using a complete crosstalk noise model," *Proceedings 2002 Design, Automation and Test in Europe Conference and Exhibition*, Paris, France, 2002, pp. 456–463.
14. A. Tada et al. "Delayed-ABC SOI for crosstalk noise repair", *IEICE Electronic Express*, vol. 5, pp. 354–360, 2008.
15. M. Hashimoto and H. Onodera, Crosstalk Noise Optimization by Post-Layout Transistor Sizing, *IEICE Trans. on Fundamentals of Electronics, Communications and Computer Sciences*, vol. E87-A, no. 12, pp. 3251–3257, 2004.
16. P. Heydari and M. Pedram, Capacitive Coupling Noise in High-Speed VLSI Circuits, IEEE Trans. on Computer Aided Design, no. 3, vol. 24, pp. 478–488, March 2005.
17. S. Sayil, S. Lamichhane and K. Sayil, "Crosstalk noise mitigation using a transmission gate with varied gate bias", *Analog Integrated Circuits and Signal Processing*, vol. 105, pp. 183–190, 2020.
18. S. Sayil, Soft Error Mechanisms, Modeling and Mitigation, Springer, 2016.
19. V. Ferlet-Cavrois, L. W. Massengill, and P. Gouker, "Single event transients in digital CMOS—a review," *IEEE Trans. on Nuclear Science*, vol. 60, no. 3, pp. 1767–1790, Jun. 2013.

20. J. Kumar, M.B. Tahoori, "Use of pass-transistor logic to minimize the impact of soft errors in combinational circuits", In *WKSP on SELSE*, pp. 67–74, 2005.
21. Predictive Technology Model (PTM), http://www.eas.asu.edu/~ptm, 2021.
22. T. C. Carusone, D. A. Johns, K. W. Martin, Analog Integrated Circuit Design, 2nd Edition, Wiley, 2011.

Modeling and Prediction of Crosstalk Noise

3

3.1 Introduction

Advancements in lithographic process cause circuits to become increasingly more crowded leading continuous reduction of distances among interconnects. The interaction caused by parasitic coupling between wires, usually known as crosstalk, may cause undesired effects such as positive and negative glitches, overshoot, undershoot, signal delays or even delay reduction. If crosstalk effects on the victim net are large, they can easily propagate into storage elements that connected to victim line and can cause permanent errors.

Inductance effects in on-chip interconnects becomes increasingly important with smaller rise times and lower wire resistance especially in global interconnects. However for most on-chip lines or interconnects, capacitive effects are still the dominant factor therefore inductance can be safely ignored.

In the past, SPICE based simulators have been traditionally used to calculate crosstalk noise in signal lines. Due to density of interconnect lines, these simulations are time inefficient. Therefore, a rapid and accurate crosstalk noise estimation alternative is necessary to ensure signal integrity in a limited design cycle time.

Several papers published modeled crosstalk effects using simple lumped circuit models. Vittal and Marek-Sadowska [1] modeled each aggressor and victim net by a simple L-type lumped RC circuit and obtained a crosstalk noise bound using a step input. Later, extensions to his model are made by [2, 3] to consider a saturated ramp input or a π-type lumped RC circuit. Cong later proposed a 2-π model [4] that offered a better accuracy than previous models. In this model, the victim line is modeled using the 2-π model while aggressor net is simplified as a saturated ramp at the coupling node. Here, in this model, the distributed aggressor net characteristics and its driver information have been included into the ramp slew rate at the coupling location and hence requires some sort

S. Sayil, *Noise Contamination in Nanoscale VLSI Circuits*, Synthesis Lectures on Digital Circuits & Systems, https://doi.org/10.1007/978-3-031-12751-9_3

of pre-processing on the aggressor line. Later, modifications are made to this model and an improved 4-π model has been proposed [5]. In this improved version, the model is extended to include the aggressor distributed line characteristics. However, the approach uses decoupling and during the decoupling, it ignores victim loading effect on aggressor coupling node. In addition, the non-switching (passive) aggressors coupled to victim line were simply represented as grounded capacitors at the coupling point. In real case, due to resistive shielding effect the loading effect should be less. The aggressor tree branches and multi-aggressor case were also not discussed.

The methodology to be presented here addresses all these deficiencies. This method proposes a complete multi-line model based on the 4-π model and models non-switching aggressors correctly compared to earlier work [6]. Being different than previous 4-π model, passive aggressors are represented as equivalent capacitances to the victim rather than simple lumped coupling capacitances. The equivalent capacitance formulas are derived by noting realistic exponential aggressor waveform and resistive shielding impact. Since, active aggressors are more closely approximated by exponential waveforms rather than step or ramp waveforms [7–9]; realistic exponential waveforms have been used for switching aggressors in closed-form derivations. Similarly, aggressor tree branches have also been formulated.

Another shortcoming of previous work in [5] is the too much reliance on the dominant pole approximation. In their work, the direct transfer function between aggressor input and victim output is first calculated, then dominant pole approximation is used over the whole transfer function to reduce complexity. However, this results in accuracy loss in noise prediction. In the proposed methodology, the use of dominant pole approximation has been limited to only a small portion of transfer function. In order to achieve this, the new exponential waveform at the coupling node is first estimated using a simple formulation, and then is applied to the transfer function between coupling node and the victim output to obtain noise voltage [6]. The dominant pole approach has been used only partially for the transfer function between aggressor coupling node and victim output. This achieves better results compared to previous methodology given in [5].

In this chapter, based on the model, first the aggressor coupling node waveform has been derived. After calculating the transfer function between aggressor coupling node and victim receiver, victim noise waveform is obtained. Since, noise peak and width are the two parameters that determine whether the noise is acceptable or not, the closed form analytic expressions for peak noise and noise width are also formulated. The results for 1000 random test circuits are then compared against simulations. Simulations are also done for multiple switching aggressors and results show good agreement to HSPICE results.

3.2   The 4-π Model

In the 4-π model, both victim and the aggressor net are modeled using the 2-π circuits [5]. Finally the template circuit shown in Fig. 3.1 is obtained. In this model, effective resistances R_d and R_{th} model the victim and aggressor driver resistances, respectively. Drivers are represented by linear resistors using the method described in [10]. Other RC parameters are calculated based on technology and the geometric information from Fig. 3.1a. The coupling node (node 2) is set to be the center of the coupling portion of the victim net. It is assumed that the upstream and downstream resistance/capacitance at node 2 are R_{1a}/C_{ua} and R_{2a}/C_{da}, respectively. Similarly for victim net, let's assume upstream and downstream resistance/capacitance at node 5 to be R_{1v}/C_{uv} and R_{2v}/C_{dv} respectively. Then, for aggressor and the victim, we have:

$$C_{1a} = C_{ua}/2, \quad C_{2a} = (C_{ua} + C_{da})/2 \quad \text{and,} \quad C_{la} = C_{da}/2 + C_{lda}$$

$$C_{1v} = C_{uv}/2, \quad C_{2v} = (C_{uv} + C_{dv})/2 \quad \text{and,} \quad C_{lv} = C_{dv}/2 + C_{ldv}$$

where,

C_{lda} and C_{ldv} represent the load capacitances for aggressor and victim lines, respectively.

3.3   Passive Aggressor Modeling

A victim can be coupled to many non-switching or passive aggressors. Previously the loading effect of a passive aggressor is simply taken as a coupling capacitor at victim coupling point [4, 5]. However, a passive aggressor follows victim waveform and contributes to the stability of the victim line. Therefore, equivalent load capacitance at the victim coupling point is less than coupling capacitance and can be formulated using coupling/branching admittance concept [6, 11]. The equivalent capacitance formula for a passive aggressor is derived for an exponential aggressor waveform. For this, the passive aggressor is first reduced to the simple circuit shown in Fig. 3.2b, where:

$$R_a' = R_{th} + R_{1a} \tag{3.1}$$

$$C_a' = C_{2a} + C_{la} + \left(R_{th}^2/(R_{th} + R_{1a})^2\right)C_{1a} \tag{3.2}$$

Then, the victim waveform is assumed to be a normalized exponential voltage for matching purposes as shown in Fig. 3.3. The equivalent capacitance for the passive aggressor can now be formulated. The currents coming from victim node should be same both cases:

Fig. 3.1 4-π model for two
coupled interconnects

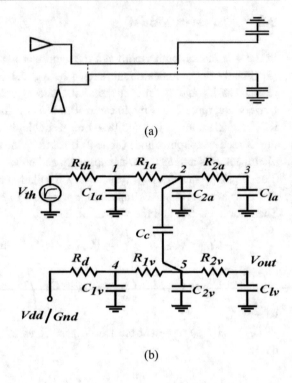

(a)

(b)

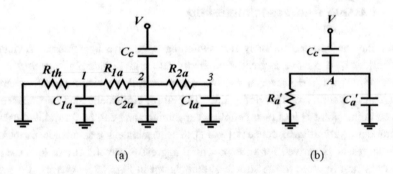

(a)   (b)

Fig. 3.2 A non-switching aggressor net coupled to the victim line

$$I = C_c \left[\frac{dV_V(t)}{dt} - \frac{dV_A(t)}{dt} \right] = C_{eq} \frac{dV_V(t)}{dt} \tag{3.3}$$

Assuming zero initial condition and an exponential waveform, the equivalent capacitance can be calculated by integrating (3.3) over $0 \leq t \leq 5t_r$ interval (t_r is the exponential
rise time constant):

$$C_{eq} = C_c[1 - V_A(5t_r)] \tag{3.4}$$

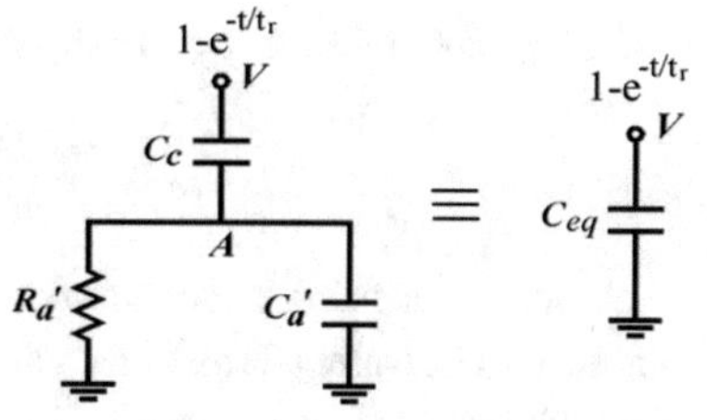

Fig. 3.3 Passive aggressor line reduction to an equivalent capacitance

Then, considering the left part of Fig. 3.3,

$$C_c \left[\frac{dV_V(t)}{dt} \right] = \left(C_c + C'_a \right) \frac{dV_A(t)}{dt} + \frac{V_A(t)}{R'_a} \tag{3.5}$$

$$V_A(t) = \frac{C_c R'_a}{t_r - R'_a (C_c + C'_a)} \left[e^{-\frac{t}{t_r}} - e^{\frac{-t}{R'_a (C_c + C'_a)}} \right] \tag{3.6}$$

Inserting $V_A(5t_r)$ in Eq. (3.4):

$$C_{eq} = C_c \left[1 + \frac{C_c R'_a}{t_r - R'_a (C'_a + C_c)} \cdot e^{\frac{-5t_r}{R'_a (C'_a + C_c)}} \right] \tag{3.7}$$

A passive aggressor coupled to the victim line can be reduced to an equivalent capacitor using the formula derived above and this capacitor would then be taken in parallel with C_{2v} at node 5. Finally, the circuit would reduce to the one in Fig. 3.1.

3.4   RC Trees and Branch Modeling

Previous work in [5] treats aggressor net branches simply as lumped capacitances at the branching point. However, the capacitance seen at the branching node is less than the total branch capacitance due to resistive shielding effect, and the approach in [5] is incorrect. Here, an equivalent capacitance formula for tree branches is derived noting the exponential aggressor waveform [6].

First tree branches are reduced to a simple π model following the moment matching method as demonstrated in [11]. Then, this model reduces to an equivalent branching capacitance $C_{eq\text{-}br}$ (Fig. 3.4) considering an exponential waveform on input node A.

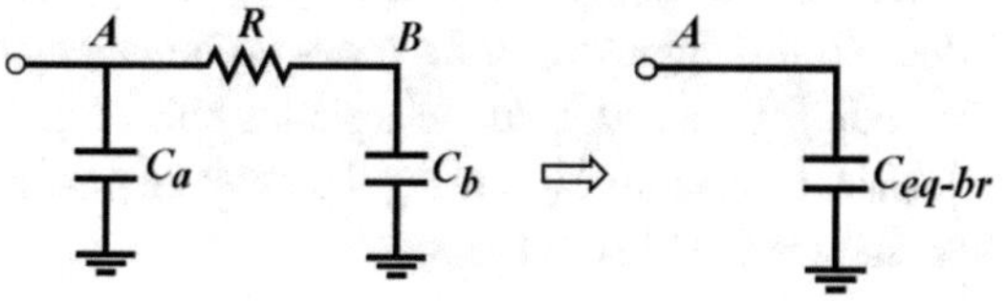

Fig. 3.4 Obtaining an equivalent branching capacitance

In Fig. 3.4, one can equate the currents on node A for both circuits:

$$C_{eq-br} \frac{dV_A(t)}{dt} = C_a \frac{dV_A(t)}{dt} + C_b \frac{dV_B(t)}{dt} \tag{3.8}$$

Assuming a rising exponential voltage on input node and zero initial condition, we can obtain an equivalent branching capacitance by integrating both sides of above equation over $0 \leq t \leq 5t_r$ time interval:

$$C_{eq-br} = C_a + C_b V_B(5t_r) \tag{3.9}$$

Applying KCL on node B, one obtains:

$$\frac{dV_B(t)}{dt} + \frac{V_B(t)}{RC_b} = \frac{1 - e^{-t/t_r}}{RC_b} \tag{3.10}$$

After solving the differential equation and taking $t = 5t_r$,

$$V_B(5t_r) = 1 + \frac{RC_b}{t_r - RC_b} \cdot e^{-5t_r/RC_b} \tag{3.11}$$

Finally, this value can be inserted in (3.9) giving:

$$C_{eq-br} = C_a + C_b \left[1 + \frac{RC_b}{t_r - RC_b} e^{-5t_r/RC_b} \right] \tag{3.12}$$

3.5   Aggressor Waveform Calculation at Coupling Node

In comparison to the approach in [5], the proposed model uses a reduced transfer function between aggressor coupling node and the victim node, hence results in small accuracy loss. In previous work, the direct transfer function between aggressor input and victim output is first calculated, then dominant pole approximation is hired over the whole transfer function to reduce complexity. However, too much use of dominant pole approximation always causes loss in model accuracy.

In the proposed model, the aggressor waveform at the coupling node is calculated and then entered to the transfer function between the coupling node and the victim output to obtain victim noise voltage. The moderate use of the dominant pole approximation results in increased model accuracy.

In order to model the coupling node aggressor waveform correctly, victim-loading effect on the aggressor node needs to be calculated. The loading effect is smaller than the coupling capacitor due to resistive shielding. The victim line can be reduced into an equivalent capacitor C_{eqv} using the quiet aggressor/victim net reduction techniques which was summarized in Sect. 3.3.

Fig. 3.5 Tree branch reduction on right part of aggressor net

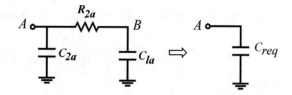

The aggressor branches after the coupling point are also reduced to an equivalent capacitance C_{req} using the tree branch reduction techniques given earlier. Here, first consider the right part of the aggressor net after the coupling point which is a π-type circuit (Fig. 3.5).

The equivalent branching capacitance C_{req} is similarly given by (shown here for convenience):

$$C_{req} = C_{2a} + C_{la}\left[1 + \frac{R_{2a}C_{la}}{t_r - R_{2a}C_{la}}e^{\frac{-5t_r}{R_{2a}C_{la}}}\right] \tag{3.13}$$

After application of the reduction techniques, the 4-π network shown in Fig. 3.1b reduces to Fig. 3.6 for aggressor coupling node voltage calculation. From Fig. 3.6,

$$V_1(s) = \frac{Z}{R_{th} + Z}V_{in}(s) \tag{3.14}$$

where,

$$\frac{1}{Z} = \frac{1}{R_{1a} + \frac{1}{sC}} + sC_{la} \tag{3.15}$$

and,

$$C = C_{2a} + C_{req} + C_{eqv},$$

then, we have

$$V_2(s) = \frac{1}{sCR_{1a} + 1}V_1(s) \tag{3.16}$$

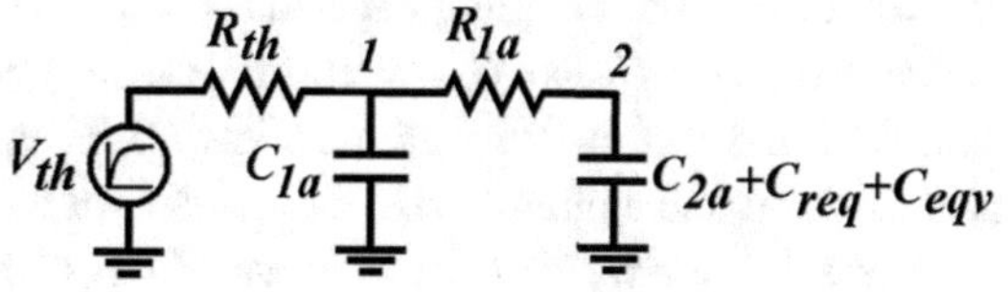

Fig. 3.6 The circuit for aggressor waveform calculation at coupling node

Finally, the transfer function between the input and coupling node 2 is:

$$\frac{V_2(s)}{V_{in}(s)} = \frac{1}{st_a + 1} \tag{3.17}$$

where,

$$t_a = C_{1a}R_{th} + (C_{2a} + C_{veff} + C_{reff})(R_{th} + R_{1a})$$

We note that t_a in fact the Elmore delay between the input and Node 2. For unit saturated exponential input $V_{in}(t) = 1 - e^{-t/t_r}$, we obtain:

$$V_2(t) = 1 + \frac{1}{t_r - t_a}\left(t_a e^{-t/t_a} - t_r e^{-t/t_r}\right) \tag{3.18}$$

This waveform, when plotted, represents a delayed exponential waveform as expected. However, this needs to be reduced to only one exponential term for simplicity. If we assume the delayed waveform at coupling node to be

$$V_2(t) = 1 - e^{-t/t_x} \tag{3.19}$$

We equate these two waveforms:

$$1 + \frac{t_a}{t_r - t_a}e^{-t/t_a} - \frac{t_r}{t_r - t_a}e^{-t/t_r} = 1 - e^{-t/t_x}$$

The area under both exponential functions should be same:

$$\int_0^\infty \frac{t_a}{t_r - t_a}e^{-t/t_a} - \int_0^\infty \frac{t_r}{t_r - t_a}e^{-t/t_r} = \int_0^\infty -e^{-t/t_x}$$

It turns out that:

$$t_x = t_r + t_a \tag{3.20}$$

The new exponential time constant t_x calculated has been verified by plotting the function in (3.19) simultaneously with HSPICE result. The following parameter values were used for the verification: $R_{th} = 200\ \Omega$, $R_{1a} = R_{2a} = 120\ \Omega$, $R_d = 250\ \Omega$, $R_{1v} = R_{2v} = 100\ \Omega$. The coupling capacitance C_c is taken as 150 fF. Other capacitances are given as follows: $C_{ua} = C_{da} = 100$ fF, $C_{uv} = C_{dv} = 100$ fF. The load capacitances for aggressor and victim line were taken as 50 fF each. A normalized aggressor voltage is also assumed and aggressor rise time t_r chosen as 150 ps.

Figure 3.7 shows the calculated waveform (3.19) versus HSPICE result at aggressor coupling node. Although some mismatch exists especially before the waveform saturates, the time corresponding to 50% point shows good matching property.

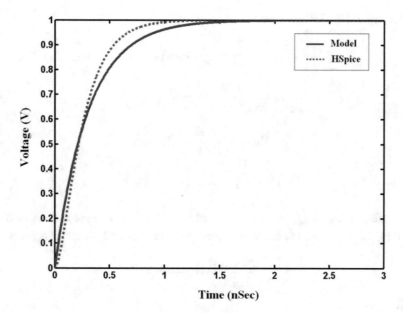

Fig. 3.7 Coupling point waveform of the aggressor

For the given parameter values, the model predicts the new rise time constant (t_x) as 305 ps while HSPICE calculates it as 314 ps resulting in an error of only 2.8%. For 100 random circuits, the model has been verified and error corresponding to each case is calculated and it has been found the absolute error value remains less than 3.5%.

3.6   Output Voltage Formulation

In the previous section, the aggressor waveform at the coupling node is formulated by considering the exponential aggressor input. Now, the aggressor waveform at coupling location needs to be entered to the transfer function to calculate noise as shown in Fig. 3.8.

Fig. 3.8 The output voltage calculation

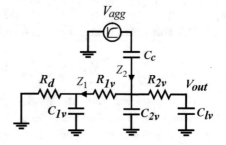

Referring to Fig. 3.8, we have,

$$\frac{1}{Z_1} = \frac{1}{R_d} + sC_{1v}$$
(3.21)

$$\frac{1}{Z_2} = \frac{1}{R_{2v} + sC_{lv}} + sC_{2v} + \frac{1}{Z_1 + R_{1v}}$$
(3.22)

$$V_2(s) = \frac{Z_2}{Z_2 + \frac{1}{sC_c}} V_{agg}(s)$$
(3.23)

$$V_{noise}(s) = \frac{1}{sC_{lv}R_{2v} + 1} V_2(s)$$
(3.24)

After substituting (3.23) in $V_{noise}(s)$ equation in (3.24), Dominant-Pole approximation method [4, 12, 13] is used to reduce complexity of the transfer function. Finally, we have:

$$\frac{V_{noise}(s)}{V_{agg}(s)} = \frac{s\tau_x}{\tau_v s + 1}$$
(3.25)

where,

$$\tau_v = R_d(C_c + C_{1v} + C_{2v} + C_{lv}) + R_{1v}(C_c + C_{2v} + C_{lv}) + R_{2v}C_{lv}$$

and

$$\tau_x = (R_d + R_{1v})C_c$$

If we insert the exponential function in (3.18) as the aggressor input voltage, we obtain the following noise waveform:

$$V_{noise}(t) = \frac{\tau_x}{t_x - \tau_v} \cdot \left(e^{-t/t_x} - e^{-t/\tau_v} \right)$$
(3.26)

By differentiating $V_{noise}(t)$ with respect to t, the time the voltage peaks t_{peak} can be found:

$$t_{peak} = \frac{t_x \tau_v}{t_x - \tau_v} \log_n(t_x/\tau_v)$$
(3.27)

The noise peak voltage V_{peak} is found after inserting (3.27) in (3.26):

$$V_{peak} = \frac{\tau_x}{\tau_v - t_x} \left[(t_x/\tau_v)^{\frac{t_x}{\tau_v - t_x}} - (t_x/\tau_v)^{\frac{\tau_v}{\tau_v - t_x}} \right]$$
(3.28)

Noise peak has been traditionally used as a metric to determine if the noise is at an acceptable level. The noise width is also necessary in determining whether a noise pulse can go through a receiver. If noise peak exceeds the threshold but it does not carry sufficient width, the noise may not be received at the receiver output at all. Therefore, noise width should be considered.

The noise peak expression was already derived in (3.27). For noise width, the threshold is usually taken as 50% of V_{peak}. Considering Eq. (3.26) and the threshold, one can obtain a function $f(t)$ which can be used in Newton's iteration method to solve for the t_1 and t_2 time instances:

$$\frac{\tau_x}{\tau_v - t_x} \cdot \left(e^{-t/\tau_v} - e^{-t/t_x}\right) = \frac{\tau_x}{2(\tau_v - t_x)}\left[(t_x/\tau_v)^{\frac{t_x}{\tau_v - t_x}} - (t_x/\tau_v)^{\frac{\tau_v}{\tau_v - t_x}}\right] \quad (3.29)$$

where,

$$f(t) = e^{-t/\tau_v} - e^{-t/t_x} - 0.5(t_x/\tau_v)^{t_x/\tau_v - t_x} + 0.5(t_x/\tau_v)^{\tau_v/\tau_v - t_x} = 0$$

and

$$f'(t) = e^{-t/\tau_v}/\tau_v - e^{-t/t_x}/t_x$$

This method converges very rapidly if the initial guesses are taken carefully. The initial guesses of t_1 and t_2 are taken as $\frac{1}{4} * t_{peak}$ and $4 * t_{peak}$ respectively. The values of t_1 and t_2 are updated using the iteration formula given below:

$$t_{1_{k+1}} = t_{1_k} - \frac{f(t_{1_k})}{f'(t_{1_k})} \quad \text{and} \quad t_{2_{k+1}} = t_{2_k} - \frac{f(t_{2_k})}{f'(t_{2_k})} \quad (3.30)$$

Then, the noise width is defined by:

$$t_{width} = t_2 - t_1 \quad (3.31)$$

The algorithm converges rapidly after only a few iterations.

Model Summary
The following steps summarize for the proposed model:

1. For each passive aggressor, calculate the equivalent capacitance value C_{eq-a} at victim coupling point. This capacitor is then placed in parallel at victim line coupling point.
2. Repeat the same procedure in step (a) for the victim line, and find the equivalent capacitance of the victim line C_{eq-v}. This capacitor updates the value of C_{2a} at the active aggressor coupling point. The formula for C_{eq-v} is slight variation of Eq. (3.7).
3. Any aggressor tree branches are also reduced to an equivalent branching capacitance C_{eq-br} at branching point.
4. Calculate the aggressor coupling node waveform. Find the new rise time constant t_x of the exponential waveform.
5. Calculate other time constants τ_x, and τ_v.
6. Obtain noise peak value V_{peak} and the time it occurs, which is t_{peak}.

Table 3.1 The percentage of nets that fall into the error ranges

Error range	V_{peak} (%)	t_{width} (%)
Within ±5%	87.4	74.9
Within ±10%	97.9	86.3
Within ±15%	99.3	93.1
Avg. error	4.1	6.8

7. Use $t_{peak}/4$ and $4t_{peak}$ as initial guesses in the iteration formula derived for the width, and calculate the corresponding time instances t_1 and t_2 values to determine the noise pulse width, namely t_{width} by (3.31).

3.7   Validation of the Proposed Model

The model has been tested extensively and its accuracy has been compared against SPICE simulation results. 1000 random test circuits having 6-node template circuit topology have been generated to simulate real world cases using 130 nm technology parameters. The parameter ranges were taken as follows [5]: R_d and R_{th} are 10–1500 Ω; load capacitances for victim and aggressor lines are 5–50 fF; aggressor and victim wire resistances are 10–250 Ω; and finally t_r is chosen in the range between 20 and 500 ps.

Table 3.1 shows the percentage of nets that fall into different error ranges. For example, 97.9% of nets have errors less than 10% when predicting the noise peak voltage. Results show an average error of 4.1% for the noise peak and an average error of 6.8% for the noise width.

3.8   Multiple Active Aggressors

In a real circuit, a given victim line can be coupled to many switching (active) aggressors, and Superposition theorem can be hired to calculate the total cross-coupling noise. With Superposition, each active aggressor is switched one at a time while holding other aggressor drivers non-switching. The noise contributions are summed at the end to calculate total noise at the victim end.

If there are N switching aggressors, it is necessary to calculate noise N times to obtain final result, hence time complexity is linear. Prior to any noise calculation, an equivalent capacitance value should be calculated for each aggressor using (3.7).

The equivalent capacitance values are utilized during Superposition to represent non-switching aggressors and this reduces the complex multiline network into a manageable 4-π template shown in Fig. 3.1 during each Superposition step.

In order to further validate the results, the design data for a real microprocessor design in 130 nm technology [5] have been used and 30 noise-prone nets have been obtained. Table 3.2 illustrates experimental results obtained for multiple aggressor case. Various interconnect length, width and spacing are examined. Interconnect lengths varied from 100 up to 2000 μm. Some of these nets also included tree branches.

Experiments are performed up to eight aggressors. The second column in the table specifies the number of switching aggressors and the total number of aggressors. In these experiments, different switching windows of aggressors have also been considered.

In Table 3.2, noise peak and width values of the previous approach in [5], and the proposed approach have been compared with HSPICE results. HSPICE simulation utilized a 20-π representation for each wire along with distributed coupling capacitance. The proposed approach has an average error of only 4.7% for the noise peak, and 7.1% for the noise width. In comparison, Becer's approach [5] has an average error of 10.7% for the noise peak, and 11.9% for the noise width. The inclusion of victim loading effect, the equivalent capacitance representation for passive aggressors and moderate use of dominant pole approximation makes the model superior in terms of accuracy. On the other hand, the CPU time for these 30 industrial nets ranges from 0.02 to 0.15 mS on a Pentium machine.

3.9   Conclusion

This chapter proposed an accurate method to estimate crosstalk noise in the presence of multiple aggressor lines. The proposed model presented a complete multiline noise model by representing active and passive aggressors simultaneously. For passive aggressors, an equivalent capacitance model has been derived noting realistic exponential aggressor waveform and resistive shielding effects. General closed-form expressions for cross-talk noise peak and width have been developed. Closed form expressions for noise width and height are compared against HSPICE and results are very promising. Results show that the average error for noise peak is 4.1% and for the width is 6.8% while allowing for very fast analysis. The model has also been verified for multi aggressor scenario. The proposed model can be used to allow time-efficient accurate analysis of multi conductor line structures.

Table 3.2 Experimental results obtained for industrial circuits

Case #	Act/total	Noise peak (HSPICE) mV	(Becer) mV	err%	(Model) mV	err%	Noise width (HSPICE) nS	(Becer) nS	err%	(Model) nS	err%
1	2/2	252	219	13.09	258	2.38	1.28	1.46	14.06	1.33	3.91
2	3/5	340	315	7.35	354	4.12	1.46	1.57	7.53	1.38	5.48
3	3/5	384	345	10.15	391	1.82	1.29	1.37	6.20	1.35	4.65
4	3/5	390	358	8.21	371	4.87	1.58	1.71	8.23	1.44	8.86
5	4/5	492	438	10.98	470	4.47	1.64	1.80	9.76	1.51	7.93
6	3/3	331	305	7.85	354	6.95	1.47	1.66	12.93	1.46	0.68
7	2/3	220	197	10.45	203	7.73	1.18	1.29	9.32	1.24	5.08
8	2/2	296	259	12.50	272	8.11	1.12	1.28	14.28	1.21	8.04
9	5/6	394	364	7.61	384	2.54	1.19	1.41	18.48	1.26	5.88
10	4/5	125	108	13.60	118	5.60	2.01	2.23	10.94	1.85	7.96
11	6/6	470	417	11.28	453	3.62	1.71	1.88	9.94	1.82	6.43
12	6/6	360	314	12.78	383	6.39	1.66	1.85	11.45	1.77	6.63
13	8/8	510	449	11.96	532	4.31	1.59	1.77	11.32	1.47	7.55
14	5/8	490	447	8.78	477	2.65	1.09	1.17	7.34	0.99	9.17
15	4/4	526	468	11.03	512	2.66	1.89	2.01	6.35	2.03	7.41
16	3/3	183	169	7.65	177	3.28	0.95	1.15	21.05	0.83	12.63
17	4/4	372	344	7.53	383	2.96	1.13	1.25	10.62	1.01	10.62
18	2/2	417	365	12.47	402	3.60	1.38	1.54	11.59	1.41	2.17
19	7/8	517	440	14.89	477	7.74	1.44	1.61	11.81	1.38	4.17
20	6/8	550	478	13.09	588	6.91	2.38	2.54	6.72	2.43	2.1

(continued)

Table 3.2 (continued)

Case #	Act/total	Noise peak (HSPICE) mV	(Becer) mV	err%	(Model) mV	err%	Noise width (HSPICE) nS	(Becer) nS	err%	(Model) nS	err%
21	5/8	523	479	8.41	539	3.06	1.79	1.93	7.82	1.72	3.91
22	4/7	411	365	11.19	393	4.38	2.14	2.41	12.61	2.28	6.54
23	5/6	464	419	9.70	433	6.68	0.87	0.98	13.79	0.79	9.2
24	4/5	412	369	10.44	388	5.83	1.24	1.39	12.10	1.13	8.87
25	6/6	510	465	8.82	481	5.69	1.34	1.53	14.18	1.22	8.96
26	3/4	282	249	11.70	269	4.61	0.79	0.91	15.19	0.69	12.66
27	4/5	324	289	10.80	314	3.09	1.55	1.74	12.26	1.41	9.03
28	5/5	482	441	8.51	452	6.22	1.01	1.24	19.80	0.96	4.95
29	3/3	212	171	19.34	227	7.08	0.89	1.05	17.97	0.78	12.36
30	5/6	395	361	8.61	389	1.52	2.12	2.33	9.91	1.96	7.55
Average error (%)				10.7		4.7			11.9		7.05

References

1. A. Vittal and M. Marek-Sadowska. "Crosstalk reduction for VLSI", *IEEE Transactions on Computer-Aided Design*, Vol. 16, pp. 1817–1824, 1997.
2. A. B. Kahng, S. Muddu, and D. Vidhani, "Noise and delay uncertainty studies for coupled RC interconnections", *IEEE Int. ASIC/SOC Conf.*, pp. 3–8, 1999.
3. S. Nakagawa, D. M. Sylvester, J. McBride, and S.-Y. Oh "On-chip crosstalk noise model for deep submicrometer ULSI interconnect", *H. P. Journal*, Vol. 49, pp. 39–45, 1998.
4. J. Cong, D.Z. Pan, and P. V. Srinavas, "Improved crosstalk modeling for noise constrained interconnect optimization", *Proceedings of ASP/DAC*, pp. 373–378, 2001.
5. M. R. Becer, D. Blaauw, V. Zolotov, R. Panda, I. N. Hajj, "Analysis of noise avoidance techniques in DSM interconnects using a complete crosstalk noise model", *2002 Design, Automation and Test in Europe Conference*, pp. 456–464, 2002.
6. Sayıl, S., Borra, U.K. A multiline model for time-efficient estimation of crosstalk. *Analog Integr Circ Sig Process* 59, 65–75 (2009).
7. T. Sato, Y. Cao, K. Agarwal, D. Sylvester, and C. Hu, "Bidirectional closed-form transformation between on-chip coupling noise waveforms and interconnect delay change curves," *IEEE Transactions on Computer-Aided Design*, vol. 22, no. 5, pp. 560–572, 2003.
8. K. Agarwal, Y. Cao, T. Sato, D. Sylvester, and C. Hu, "Efficient generation of delay change curves for noise-aware static timing analysis" *Proc. of Asia and South Pac. Design Automation Conf.*, pp. 77–84, 2002.
9. M. Kuhlmann, S. S. Sapatnekar, K. K. Parhi, "Efficient crosstalk estimation", *International Conference on Computer Design (ICCD '99)*, pp. 266–272, 1999.
10. R. Levy, D. Blaauw, G. Braca, A. Dasgupta, A. Grinshpon, C. Oh, B. Orshav, S. Sirichotiyakul, and V. Zolotov, "Clarinet: A noise analysis tool for deep submicron design", in *Proc. Int. Conf. Computer-Aided Design*, Nov. 2002, pp. 587–594.
11. J. Qian, S. Pullela, and L. T. Pillage, "Modeling the effective capacitance for the RC interconnect of CMOS gates," *IEEE Transactions on Computer-Aided Design*, Vol. 13, pp. 1526–1535, 1994.
12. L. T. Pillage and R. A. Rohrer, "Asymptotic Waveform Evaluation for Timing Analysis", *IEEE Transactions on Computer-Aided Design*, Vol. 9, No. 4, pp. 352–366, 1990.
13. E. Acar, A. Odabasioglu, M. Celik, and L. Pileggi. "S2p: a stable 2-pole RC delay and coupling noise metric IC interconnects", *Proceedings 9th Great Lakes Symposium on VLSI*, pp 60–63, 1999.

Clock Uncertainty and Power Supply Noise

4

4.1   Clock Skew and Jitter

The design of clock distribution networks in synchronous digital systems presents pose greater challenges. Controlling the clock signal delay and transition time in the presence of various noise sources, process variations, and thermal effects pose a greater challenge in the design of synchronous circuits. Here, we first introduce some basic concepts relating to clock network design.

Clock Skew

Clock skew occurs when different parts of the circuit receive the clock signal at different times. It is defined as the variation at clock arrival times at two different destination points in the clock network such as the inputs of register elements [1]. In the presence of clock skew, one storage element might use another's output before it is ready resulting in unpredictable results. Ideally, we want clock skew to be zero meaning it arrives at all points on an IC at the same time. In real application, clocks have uncertainty in their arrival times. Skew can be positive or negative. When data and clock are routed in same direction, a positive skew occurs. Figure 4.1 shows a positive skew case.

While a positive skew is beneficial to improve the set-up time, it may cause a hold time violation. A negative skew, on the other hand, can lead to setup time violation while improving the hold time violation. In the negative skew case, data and clock signal are routed in the opposite direction as seen in Fig. 4.2.

Clock skew is caused by the mismatches along the clock path and the load differences. Figure 4.3 shows a case where different clock skews occur due to wire length difference. Since longer wire lengths translate into larger delays, the register elements far from the clock source suffers the largest skew.

© The Author(s), under exclusive license to Springer Nature Switzerland AG 2022   63
S. Sayil, *Noise Contamination in Nanoscale VLSI Circuits*, Synthesis Lectures on Digital
Circuits & Systems, https://doi.org/10.1007/978-3-031-12751-9_4

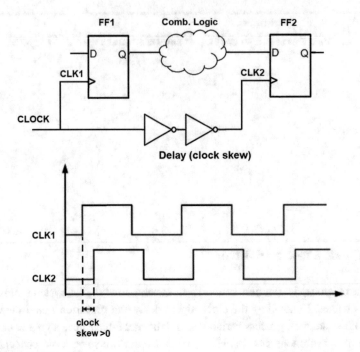

Fig. 4.1 Positive clock skew

Clock Jitter

Clock jitter can be defined as the variation of the clock edge from one cycle to another. As a result, the clock period can basically vary based on the current cycle which causes temporal uncertainty. As an example, a signal transitioning occurring on a neighboring wire can induce crosstalk noise delay on a clock network which can translate into a clock jitter. The jitter value can be positive or negative. Jitter can also emanate from internal circuitry of the phase-locked loop (PLL) which serves as the clock source (see Fig. 4.4).

4.2   Sources of Skew and Jitter

Ideally, we would like to have clock arrive at the storage elements with the same amount of delay to keep everything synchronous. However, various factors such as process and environmental variations, signal coupling from neighboring lines on a chip contribute significant delay variations in long clock lines resulting in clock skew and jitter. In the simplest sense, the wire length differences on clock signals can result in a different clock skew as shown in Fig. 4.3.

Longer wire lengths translate into larger clock skews, hence the latch elements far from the clock source suffers the largest delay. Other reasons can be capacitive loading

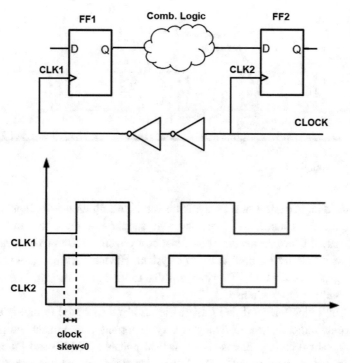

Fig. 4.2 Negative clock skew

Fig. 4.3 Various clock skews that occur due to wire length difference

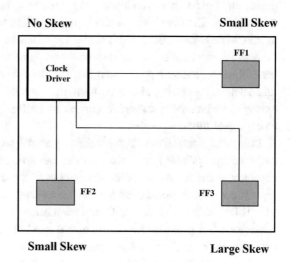

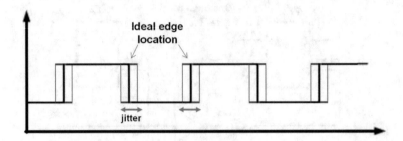

Fig. 4.4 Clock with jitter

mismatch, material imperfections, temperature variations, differences in input capacitance of latches at the clock inputs and dynamic power supply induced delay variations.

The PLL-based Clock Generator provides a cost effective solution for generating various frequencies that are required in today's system. Having said that jitter can arise due to the clock generator itself [2]. This happens due to the noise sensitive internal circuitry of the PLL, and the thermal noise.

The goal of a clock network is to distribute an identical signal to multiple locations, hence device and interconnect matching is very important. Some manufacturing process steps, most notably etching, chemical–mechanical polishing (CMP) and lithography, are influenced by the chip topography. The layout dependent processing causes systematic device and interconnect variations and results in a skew due to the random physical variations [3–5]. The transistor matching in the buffers along multiple clock paths is critical to minimizing skew effects [6]. Process variations such as oxide variations, dopant variations can impact the transistor threshold and parasitic capacitances. Hence, even for two parallel interconnects having same lengths, the delay may be different due to the buffer mismatches. Processing mismatch translates directly into loss of clock performance. For example, variations in saturation current or buffer thresholds can both lead to variable clock arrival times.

Integrated circuit fabrication processes generally result in gradients in line width (both metal and polysilicon) and thin film thickness (metal wires, interlayer dielectric). The capacitance and resistance of interconnect can vary due to changes in vertical and lateral dimensions. An important source of interconnect variation is the Inter Level Dielectric (ILD) thickness variation. Chemical–mechanical polishing at longer lengths cannot achieve perfect planarization at the global level clock wire; hence the line capacitances may vary even for the two interconnects having the same length.

Global environmental variations, specifically the temperature and DC supply voltage variations significantly contribute to skew and jitter [7, 8]. The different thermal profiles across the chip can affect the resistivity of the metal, channel mobility, and threshold voltages. The transistor thresholds strongly impacted by thermal effects. As the transistor drive currents will be affected in line with the threshold, two identical buffers hence may exhibit different delays based on the temperature.

Power supply variations, on the other hand, are a major source of clock jitter. The transistor currents are directly impacted by variations in supply. Smaller power supply will cause clock buffers to switch slower due to reduced current drive. This will cause clock edge being moved from time to time resulting in a clock jitter. Jitter here is a function of the switching activity as more gates switch in a given time, the less would be the voltage delivered to devices and hence it would induce more delays.

Clock gating is a technique that is used to save dynamic power. With clock gating, the clock to certain parts of the digital design is turned off whenever they are not needed in order to save power [9]. Clock gating, however, has major implications on clock skew and jitter. It can cause varying thermal profiles across a chip since the idle parts will have a different temperature resulting in a clock skew. The load currents can also change in between the idle and active states with clock gating. This will in turn change the power supply voltage due to resistive voltage drops in the power line resulting in a clock jitter.

There are also time-varying effects including capacitive and inductive coupling between signal and clock lines which can result in a signal-dependent capacitance. The capacitive coupling is the most dominant effect among the two. The coupling event can result in a timing uncertainty. If designers are not careful in designing the layout, the switching activity in a neighboring signal line can couple noise into the clock network. If the coupling event occurs during a time that clock signal is steady, the coupled charge can result in a glitch or a false clock pulse appearing in clock network. In this case, data racing event may occur. If the coupling or crosstalk noise is injected near the clock edge, the injected noise can move the clock edge resulting in a clock jitter.

Figure 4.5 shows an example of a clock jitter in a flip-flop (FF) configuration. Figure 4.5a shows the output of the FF when there is no clock upset, and Fig. 4.5b corresponds to the output of the same FF, with the negative crosstalk pulse affecting on the second clock pulse. In the second case, output signal "Out" is delayed by T1 due to coupling noise. Assuming the output signal "Out" is connected later to another storage element, incorrect data storage may occur if the delayed output signal arrives during the set-up time of the receiving sequential.

A crosstalk event can also cause data racing if a positive crosstalk pulse is generated on the clock network when there is no clock pulse present. If the crosstalk noise pulse generated on the clock network carries sufficient magnitude and width, it can cause incorrect latching of a data and hence an error occurs.

Figure 4.5c shows the output of flip-flop when a data racing occurs on the clock node due to a positive SET pulse appearing on clock network. Due to the formation of this pulse, data is registered early. Hence, the data arrives early by a time T2 as shown. The early latching of data may result in incorrect data storage as well.

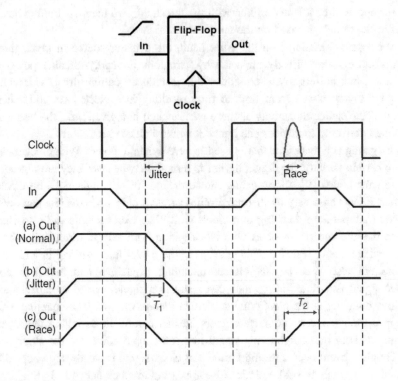

Fig. 4.5 Clock jitter and race due to a coupling event

4.3   Clock Distribution

The capacitive load variation due to load mismatches can also lead to clock uncertainty. Load variation at every stage in clock distribution may occur. For example, variation in the input capacitance of storage elements may occur or ground capacitance of interconnect lines can differ as well. Both events result in a clock jitter.

The global clock must be distributed across the chip in a way that reaches all of the clocked elements at nearly the same time. It is necessary to design a clock network that minimizes skew and jitter.

One popular scheme for clock distribution is the H-Tree which provides a well-balanced signal propagation scheme for clock distribution [1]. As can be seen from Fig. 4.6, it has been designed to minimize the delay of the clock signal from the clock source to the destination points. An H-tree consists of a global H network which distributes the clock signal to various parts and then the local H networks distribute it to the leaf nodes. The arrival times at every leaf node would be equal in this arrangement.

Fig. 4.6 H-tree clock distribution method

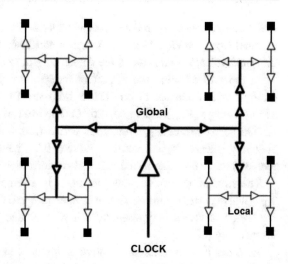

4.4 Thermal Impact on Clock Skew

In conjunction with technology scaling, chip power consumption rises, which in turn increases chip operating temperature. The increased temperature affects both clock interconnect and driving strength of clock buffers. On the other hand, the on-chip temperature gradient is a major design concern for current and future technology nodes. Clock skew will be further exacerbated by the temperature difference occurring on various parts of the clock tree. The temperature impacts the delay of a clock tree by changing the device and wire performance.

From the interconnect perspective, temperature mainly impacts interconnect resistance because the resistivity of interconnect material increases proportionally with temperature. The raised chip temperature increases the resistance of interconnect wires due to positive temperature coefficient of commonly used interconnect materials such as copper. The effect can be expressed as the following:

$$R(T) = R_0[1 + \alpha_R(T - T_0)] \tag{4.1}$$

where,

R_0 is the resistance of the interconnect at a nominal temperature,

α_R is the temperature coefficient of resistance,

T represents the local temperature of the wire.

Temperature has a significant impact on buffers, but it is not always straightforward because two important parameters, the carrier mobility and threshold voltage, have conflicting effects on transistor current when temperatures change [10]. As temperature increases, the carrier mobility decreases which causes device current to reduce. On the other hand, increased temperature decreases device threshold voltage causing drive current to increase. Depending on the supply voltage, device current may increase or decrease. At nominal supply voltage (~1 V), change in carrier mobility dominates over the change of threshold voltage which causes the device current to decrease with increased temperature.

In order to examine the impact of temperature on drain current, Fig. 4.7 plots the drain saturation current of an NMOS transistor in 45 nm technology as a function of supply voltage (V_{DD}) at 25 and 125 °C. The inset of Fig. 4.7 shows the diode connected NMOS device where V_{DD} voltage is varied from 0 to 1 V. As can be seen from figure, at supply voltage close to nominal value (~1 V), device current decreases with increased temperature change due to the fact that carrier mobility is the dominant effect.

The simulation results show that there is a temperature insensitive voltage point or V_{INS} where current is insensitive to temperature variation and the value is around 0.520 V for NMOS device. Many researchers have exploited V_{INS} region to achieve temperature insensitivity [10–12].

As a result of all this, temperature differences in different parts of clock can lead to significant clock skews. As highlighted in [10] for the 45-nm technology node, the clock skew increases in proportion to increasing temperature difference between different parts of the chip.

Researchers have proposed various schemes in order to maintain driver strength constant with respect to temperature effects so that delay uncertainty or clock skew can be

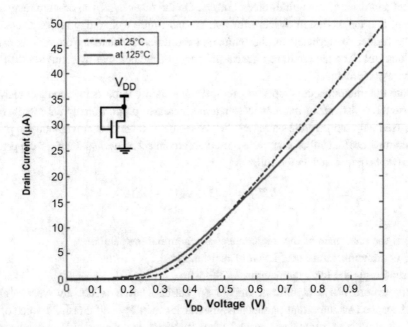

Fig. 4.7 Drain saturation current as a function of supply voltage and temperature

minimized. Chapter 7 covers these thermal compensation methods used for clock skew minimization in great detail.

4.5   Power Supply Noise

The power distribution networks are designed to provide needed voltages and currents to the devices that perform the logic functions of a chip. The power distribution system should provide a steady voltage in the presence of very large DC and transient current demands. The chip power distribution systems are normally designed to provide constant supply voltage across a chip reliably over its lifetime. However, the voltages on supply and ground lines are no longer constant and include noise component which include DC and high frequency inductive noise. Power supply noise can degrade noise margin of logic gates impacting overall timing and circuit functionality. It impacts propagation delay of logic gates since current drive capabilities of devices are affected.

Power lines are impacted by two different types of voltage drops: IR voltage drops and Ldi/dt voltage variations.

With increasing number of devices, the average current demand in an IC increases in nanoscale technologies. As fabrication technologies are getting scaled, the size of the wires has been getting smaller meaning they are getting thinner but their lengths are not getting shorter. Larger chip sizes also increase the line resistance of the power and ground distribution networks due to increased distances. This leads to an increase in resistance per unit length of the wires thereby worsening the impact of resistive power and ground noise.

When a large number of gates switch simultaneously in an IC, a significant amount of current is drawn from the power distribution network and then returns to the ground distribution network. Due to the resistance of the distribution lines, resistive voltage drops occur on power and ground lines as can be seen from Fig. 4.8. The voltage drop on the VDD line is called the "IR-drop", while its counterpart occurring on the ground network is usually referred as the "ground bounce".

IR drop is a dynamic event due to simultaneous switching events such as clocks, large buffers and drivers. During dynamic operation, the simultaneous switching activity can cause severe voltage drops on the power grid. The effect of IR drop on chip performance can be significant.

As device densities and switching currents increase, larger currents flow in the power distribution network causing larger IR drops. IR drop is dependent on the amount of circuit activity which in turn dependent on the input vector. A certain input vector may activate more functional units compared to some other vector. In other words, the demand on power line can be more and hence more IR noise will be induced.

The voltage drops occurring on these lines are made worse by the presence of Ldi/dt voltage variations occurring at package pins. The rate of change in the current increases as

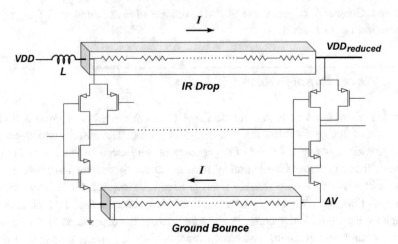

Fig. 4.8 Power grid noise

the circuit frequency rises. As a result, inductive voltage drops are now more pronounced in newer technologies. Ldi/dt voltage drops are usually referred as the high frequency noise component of the power supply noise. While the package pin inductance remained relatively constant over the years, increasing circuit frequency is responsible for increasing Ldi/dt voltage drops occurring in power grid.

Power supply noise can degrade voltage noise margins of logic gates. Over the years, power supply voltage has been scaled along with device size in order to save power and to avoid strong electric fields within devices. Referring to the high electric fields, undesirable results such as hot-electron effects, transistor punch-through effects, and device breakdown could occur. The supply voltage reduction, on the other hand, results in smaller noise margins with every new generation. With IR and Ldi/dt voltage drops, the logic gate margins are reduced even more making devices more susceptible to noise and interference.

As the power supply voltage and ground voltage level are changed due to power supply noise, the transistor currents will be impacted. The propagation delay of logic gates strongly depends on the power supply voltage level. Considering a simple CMOS inverter, the drain current of its transistors decrease as the gate to source voltage decreases. Since the source terminals of PMOS and NMOS transistors are connected directly to VDD and ground, respectively; any deviation in supply voltage and ground will be reflected in the device currents. As a result, the signal propagation delay increases as a result of reduced device currents. It will take more time for node capacitances to charge and discharge on CMOS circuit nodes. Signal propagation delay increases as a result of power supply noise and causes signal timing uncertainty.

Power supply noise can significantly affect the timing of on-chip clock. A phase-locked loop (PLL) is usually used to generate the clock signal. The PLL is very sensitive to power supply noise, and it impacts the phase of on-chip clock signal.

Power supply noise is usually budgeted at 10% meaning only a 10% reduction on VDD supply is tolerated. As for the ground voltage, it cannot rise more than 10% above 0 V. In recent technologies, the power supply voltage has gone below 1 V with scaling and supply noise became more comparable to power supply voltage. It can now easily exceed the power noise budget.

There are a couple options for reducing the impact of IR drop in a power distribution system. The simplest approach is to widen the lines that experience the largest voltage drops since increasing the width decreases the resistance (and the IR drop). However, this may not always be achievable due to constraints in the routing area. Because IR drop is mostly caused by simultaneous switching events, another strategy is to stagger the gates that are switching at the same time so that they switch at slightly different intervals. By adding delays to the signals driving the gates, device switching can be staggered to lessen peak current demands.

Another effective method is the use of decoupling capacitors between power and ground lines which would deliver an extra current during high demand situations therefore lowering the impact of IR voltage drops. The decoupling capacitors would be placed near large buffers and/or power supply pins and would provide some of the charge needed during switching events. These capacitors could easily be realized by using large NMOS transistors (see Fig. 4.9). In this case, NMOS transistor gates are tied to VDD and their source and drains tied to ground which would make a perfect capacitor. Ldi/dt effects could easily be mitigated by placing large capacitances near the power pins.

A more aggressive solution would be to use a ball-grid array, also known as solder bumps. This would allow power supply connections be present at various places within the chip. Although this approach is very effective, it would not be a cost-effective solution.

On the other hand, high current densities going through a DC wire may give rise to Electromigration (EM) effect. In other words, higher average current densities lead to undesirable damaging of metal wires due to EM. EM refers to a phenomenon whereby metallic atoms are transported by electron flow at a high current density. EM is significantly affected by current density and temperature. EM in the power grid occurs due to DC currents running in conductors over a long period of time. Wire material type

Fig. 4.9 Decoupling capacitor

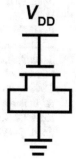

is another factor which can make a difference in terms of EM. One of the reasons for switching to copper from aluminum was due to copper's good EM property.

With increasing transistor densities and increasing chip dimensions, the current density has seen a general increase over the years. If the current density is too high, metal atoms can transport and cause permanent failures such as an open or a short failure. The metal lines should be designed wide enough to carry the average current passing thorough the wire to avoid EM.

EM failures can be reduced in several ways. The main idea in all approaches is to reduce the average current density seen by any metal segment which would require widening the metal lines. However, increasing the width beyond a certain point would be costly in terms of area and can reduce yields.

The goal of any changes to the power grid would be to decrease the probability of failure to an acceptable level. Power grid integrity can only be obtained when effects such as IR drop, ground bounce, Ldi/dt, and electromigration are considered together.

References

1. JM Rabaey, AP Chandrakasan, B Nikolić, Digital integrated circuits: a design perspective, 2nd Edition, Pearson Education, 2003.
2. D. Jeong et al., "Design of PLL-Based Clock Generation Circuits", *IEEE Journal on Solid State Circuits*, vol. SC-22, no 2, April 1987, pp 255–261.
3. Tomohisa Mizuno, Jun-ichi Okamura, and Akira Toriumi, "Experimental study of threshold voltage fluctuation due to statistical variation of channel dopant number in MOSFET's", *IEEE Tran. on Electron Devices*, vol. 41, no. 11, pp. 2216–2221, 1994.
4. Martin Eisele, Jorg Berthold, Doris Schmitt-Landsiedel, and Reinhard Mahnkopf, "The impact of intra-dive device parameter variations on path delays and on the design for yield of low voltage digital circuits", *IEEE Tran. on VLSI*, vol. 5, no. 4, pp. 360–368, 1997.
5. Xinghai Tang, Vivek K. De, and James D. Meindl, "Intrinsic MOSFET parameter fluctuations due to random dopant placement", *IEEE Tran. on VLSI*, vol. 5, no. 4, pp. 369–376, 1997.
6. Sanjay Dhar, Mark A. Franklin, and Donald F. Wann. "Reduction of clock delays in VLSI structures", *IEEE Intl. Conf. on Computer Design*, pp. 778–783, October 1984.
7. M. Hatamian, "Understanding Clock Skew in Synchronous Systems," in *Concurrent Computations*, Plenum Publishing, pp. 86–96, 1988.
8. D. Miller, "Rationale and Challenges for Optical Interconnects to Electronic Chips", *Proc. of the IEEE*, pp. 728–749, June 2000.
9. C. Mead and L. Conway, *Introduction to VLSI Design*, Addison-Wesley, 1980.
10. T. Ragheb, A. Ricketts, M. Mondal, S. Kirolos, G. M. Links, V. Narayanan, and Y. Massoud, "Design of thermally robust clock trees using dynamically adaptive clock buffers," *IEEE Trans. Circuits and Systems I*, vol. 56, no. 2, pp. 374–383, Feb. 2009.
11. A. Bellaouar, A. Fridi, M. J. Elmasry, K. Itoh, "Supply voltage scaling for temperature insensitive CMOS circuit operation", *IEEE Trans. Circuits Syst. II Express Briefs*, vol. 45, no. 3, pp. 415–417, Mar. 1998.

12. R. Kumar and V. Kursun, "Reversed temperature-dependent propagation delay characteristics in nanometer CMOS circuits", *IEEE Trans. on Circuits and Systems*, vol. 53, no. 10, pp. 1078–1082, Oct. 2006.

Substrate Noise

5

5.1 Substrate Noise Mechanisms

Complex digital circuits are often integrated with high-performance analog circuits on the same substrate. The switching transients generated by digital circuits introduce noise into the common substrate. The substrate becomes the medium that permits noise to easily propagate from digital parts into analog parts within the chip. Analog circuits do not have the wide noise margins of digital circuits, so they are susceptible to substrate voltage fluctuations. In this case, switching events generated by the digital circuits can interfere with sensitive analog components.

When current is injected into the substrate, a fluctuation in substrate voltage will occur. This voltage is usually referred as the substrate noise. The injected current can be as a result of power lines coupling noise into substrate thru ohmic contacts, wells capacitively coupling noise thru reverse biased bulk or well junctions and transistors capacitively coupling noise through the drain and source diffusions [1]. Substrate noise occurs since substrate does not have zero impedance and due to nonzero conductivity of its material a coupling is made to other devices. The substrate voltage changes occur as a result of changing device voltages or currents that are placed on the same substrate.

As mentioned above, there are three mechanisms responsible for the substrate noise:

1. Capacitive Coupling: Switching noise can be coupled into substrate through reverse biased junctions and interconnect to substrate capacitances [1, 2]. Every source and drain is capacitively coupled to the substrate through a depletion capacitance. When circuits switch, current is injected into the substrate via the source/drain junction capacitances. The amount of current injections depends on the slew rate. As switching speeds increase, the amount of coupling increases due to the displacement current formula i.e.,

i = C*dv/dt. Furthermore, the larger area of the transistors result in more noise to be coupled into the substrate due to large depletion areas.

Wires also have a ground capacitance or a capacitance to substrate. Illegitimate current pulses injected can travel through the substrate through low resistive paths. The fast switching of digital circuits make the coupling worse as the displacement current injected increases.

2. Impact Ionization: Impact ionization is caused by high electric fields between depleted parts of a drain-channel and the substrate. When electrons are accelerated by high electric fields, they collide with silicon atoms in the substrate. The electron–hole pairs created by these collisions can cause a current to flow through the substrate. In the high electric field region under the gate of an NMOS, at the pinch-off point of the NMOS transistors, accelerated electrons can create electron–hole pairs. For an NMOS device, the holes created by impact ionization flow into the substrate creating a current flowing between the drain and the substrate. Since the current flows into the ohmic substrate, it induces fluctuations in the bulk potential.

   As feature sizes are reduced, the electric field in the channel becomes stronger and therefore impact ionization currents are becoming a more important mechanism.

3. Power Supply Disturbances: Noise through power and ground rails is another source of substrate noise that needs immediate attention. The non-ideal power supplies contributes significantly to the amount of substrate noise. Power busses can couple noise into substrate thru ohmic contacts. Substrate contacts are the p+ and n+ diffusion of NMOS and PMOS transistors, respectively. It goes to either ground or VDD depending on the transistor type. The variations that occur on the power lines as a result of IR drops or Ldi/dt effects cause ringing in the power supply. The IR drop usually results from on-chip power distribution, while Ldi/dt noise results from the package or the board inductance. The IR drop can occur in both the power and ground distribution line. These resistive voltage fluctuations on the power and ground distribution networks occur due to the variations of the average current demand of the circuitry on a chip. As the substrate is connected to power and ground through low resistance substrate contacts, noise appearing on power and ground lines will directly show up on the substrate. Using decoupling capacitance can effectively reduce Ldi/dt noise and its impact.

Figure 5.1 shows the substrate noise occurring in an NMOS transistor due to capacitive coupling and power supply disturbances.

Digital circuits are not protected from substrate noise either. The noise is injected by logic gates during switching and glitch transients through impact ionization and capacitive coupling, and it is picked up by active devices via capacitive coupling and body effect [3]. As a result, the delay of the path may increase, thus possibly exceeding the predefined clock period. Such behavior is known as delay effect.

Fig. 5.1 Substrate coupling from noisy supply (left) and from source and drain junction capacitances (NMOS transistor used for demonstration)

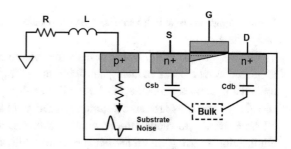

The current flow in the substrate also affects the MOS transistor by changing the effective threshold voltage value.

5.2   Minimization of Substrate Noise

Substrate noise minimization techniques include having separate analog and digital power/ground lines, creating physical separation between analog and digital circuits, using guard rings, low inductance bias path for the substrate and the use of additional substrate contacts [4].

One other method is to reduce the amount of substrate noise generated by the digital circuits. If we can reduce the switching activity, it would simply generate less substrate noise. The logic transitions can be adjusted such that gates can switch at different times. Thus, the substrate noise caused by transitions can be spread over a long period of time and hence can be minimized.

Having separate analog and digital power/ground lines is another way to reduce substrate noise. In this case digital and analog circuits and substrate all have their own dedicated power supplies. The short transients caused by analog circuits and digital circuits won't affect the substrate thorough power lines.

Having physical separation between analog and digital circuits is another method that can reduce substrate noise coupling. In this method, analog circuits are grouped together and then positioned away from digital circuits to avoid noise injection. Physical barriers include guard rings and trenches which can be positioned between digital and analog circuits.

Since much of the substrate noise generated is due to switching currents in the digital circuits, much effort has focused on designing power supply networks that inherently reduce power supply variations. As an example, substrate noise can be reduced by minimizing the pin/bond wire resistances associated with the packaging. Another way to reduce the substrate noise is to design power supply filters.

In order to reduce the junction capacitance, the bulk and the source are connected together. To eliminate the body effect the analog bulk can be tied up to ground rail. Increasing the analog ground capacitance to bulk can also be a good alternative. The higher capacitance will lower the bulk-to-source voltage transient, thereby reducing the body effect.

Materials engineering, which designs the substrate noise transmission medium, is also an effective method of reducing substrate noise. High resistivity substrates such as non-epitaxial processes attenuate substrate noise as it propagates through the substrate. Low-resistivity substrates are more commonly used and they also minimize latch-up hazards.

Another way to reduce the amount of substrate noise propagation through materials engineering is using multiple-well processes. Silicon-On-Insulator (SOI) technique can be used to minimize the noise where body is insulated from the substrate.

References

1. Ahmed Helmy, Mohammed Ismail, "Substrate Noise Coupling in RFICs", Analog Circuits and Signal Processing series, Springer, 2008 (https://doi.org/10.1007/978-1-4020-8166-8).
2. Charbon, E., Phillips, J. "Substrate Noise: Analysis, Models, and Optimization", in *VLSI: Systems on a Chip. IFIP*, vol 34. Springer, Boston, MA, 2000, https://doi.org/10.1007/978-0-387-35498-9_41.
3. S. Donnay and G. Gielen, "Substrate Noise Coupling in Mixed-Signal ASICs", Kluwer Academic Publishers, Boston, MA, 2003.
4. Peng, Mark Shane, and Hae-Seung Lee. "Study of substrate noise and techniques for minimization." *IEEE journal of solid-state circuits* 39.11 (2004): 2080–2086.

Single Event Soft Errors

6

6.1   Introduction

Terrestrial soft errors in memory have been a well-known problem [1]. However, the increasing clock frequencies and shrinking feature sizes are now making Complementary Metal–Oxide–Semiconductor (CMOS) logic more susceptible to soft errors. Reduced circuit capacitances, and power supply voltages are making CMOS circuits more vulnerable to radiation. Increasing clock frequencies can further challenge the reliability of circuits as they help to boost the error rates.

A soft error occurs when a radiation event or an incident particle induces significant charge disturbance by producing electron hole pairs to flip the data stored in a memory circuit or the output of a logic gate. The error is "soft" because the circuit or device is not permanently damaged by the radiation [2]. After a re-write process, the circuit normally resumes its operation. Due to scaling effects, circuits are becoming increasingly more sensitive to transients caused by single event (SE) particles. Mitigation of soft errors is indispensible for integrated circuits, especially in high-reliability demanding applications such as medical, satellite or space systems.

There are three primary sources of particles that can cause soft errors: alpha particles emitted from the radioactive impurities in the IC packaging, high-energy neutrons, and interaction of low-energy thermal neutrons with certain boron isotopes in the device.

The alpha particles are emitted mostly due to radioactive decay of uranium and thorium impurities located within the chip packaging. Linear Energy Transfer (LET) is the measure of energy that is transferred in the material when an ionizing particle passes through it. Most alpha particles have energies of between 3 and 7 MeV (mega or million electron-volts). Since it takes only about 3.6 eV to generate an electron–hole pair in the substrate, a 4 MeV alpha particle striking the sensitive node within a combinational logic (CL) can

© The Author(s), under exclusive license to Springer Nature Switzerland AG 2022
S. Sayil, *Noise Contamination in Nanoscale VLSI Circuits*, Synthesis Lectures on Digital
Circuits & Systems, https://doi.org/10.1007/978-3-031-12751-9_6

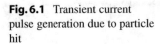

Fig. 6.1 Transient current
pulse generation due to particle
hit

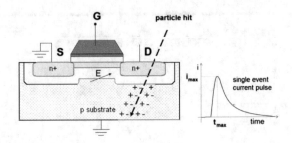

generate more than a million electron hole pairs within its particle track due to ionization
mechanism.

High-energy neutrons do not have electrical charge. While a neutron cannot induce
ionization since it does not carry any charge, secondary charged particles involved in
neutron-silicon atom collisions can generate soft errors [2]. When high energy neutrons
strike an integrated circuit, some neutrons pass through without affecting operations of
the semiconductor device, but some neutrons collide with nuclei in the silicon lattice [3,
4] resulting in the creation of secondary particles. These particles, in turn, create a trail of
electron–hole pairs. The energy of the secondary ions produced by high-energy neutron
reactions can be much higher than that of alpha particles [5].

Figure 6.1 illustrates the generation of the transient current pulse. An energetic particle
striking the sensitive area within a combinational circuit can generate large amounts of
electron–hole pairs within its ionization track.

The sensitive areas are usually the depletion regions of transistor drains or reverse-
biased p–n junctions. Under the electric field, these free carriers can drift creating a
transient current pulse which can later result in a charge collection at the struck electric
node.

The amount of the charge collected at a particular node (Q_{col}), depends on various
parameters such as device size, bias conditions, doping level, characteristics of the particle
hit, and its trajectory.

The most important electrical parameter related to soft errors is the critical charge
(Q_{crit}). At this charge, the logic state in a memory or logic circuit can be flipped. If the
collected charge (Q_{col}) is higher than Q_{crit}, the information that is stored at the circuit
node is lost. Otherwise, the data survives the upset event. Q_{crit} is related to the dynamic
response of the struck circuit, and hence the modeling of Q_{crit} is not straightforward.

The collected charge usually results in a voltage transient at the struck node. This
transient, also named as Single Event Transient (SET), can travel through combinational
logic and finally reach to a storage element if no masking effect is present. If the generated
pulse arrives at the storage element during its latching window, incorrect data will be
stored resulting in a Single Event Upset (SEU) or a soft error.

There are three masking effects that can hinder generation of a soft error. Namely, they
are electrical, logical and temporal masking. Assuming an SET transient is generated:

- If the transient generated does not carry sufficient amplitude and width, it will be filtered along the succeeding gates and attenuated at each stage. Hence, electrical masking can inhibit the SET pulse in this case.
- If the logic path the transient pulse takes is disabled by the input combination, the pulse will not propagate to the next stage. As an example, consider a 2 input NAND gate. If one of the inputs is at logic 0, the output will be automatically one. Hence, the transient pulse appearing on the other input will not be passed due to logical masking.
- When the SET pulse arrives at the latching data input, if the latching clock edge is not present SET will not be passed. Hence, temporal masking can prevent SET to cause a soft error in this case.

Figure 6.2 depicts a situation in which none of the masking effects are present. As a result, a SET results in data corruption or a soft error at the storage element's output. First, a sufficiently large transient pulse is generated at the hit site such that it propagates through a couple stages without any attenuation. Then, since the NOR gate's second input has been tied to logic 0, it allows the SET pulse travel to its output. Hence, there is no logical masking. Finally, the SET pulse is present during the rising edge of the clock pulse; the register element incorrectly stores it as valid data. This means there is no temporal masking. Finally, soft error is generated at the storage element due to absence of any masking effect.

All these masking effects are gradually diminishing with newer generation technologies. With transistors are scaled down, the node capacitances reduce, and as a result, electrical masking effects are pronounced less due to reduced critical charge. The SET pulses in newer technologies are also very strong and comparable to legitimate logic pulses. Logical masking is less effective as the logic depth in between storage elements reduces. Temporal masking is also less due to rising clock frequencies. Higher frequencies increase the likelihood of having more latching edges hence registering incorrect data.

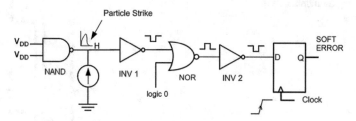

Fig. 6.2 Single event transient propagation and error generation with no masking effect present

6.2   Circuit Level Modeling of Single Event Strikes

The interaction of a charged particle with a reverse biased junction of a transistor results in a transient current pulse. The current pulse generated is usually modeled by the following double exponential waveform [1]:

$$I(t) = \frac{Q}{\tau_\alpha - \tau_\beta}\left(e^{-t/\tau_\alpha} - e^{-t/\tau_\beta}\right) \tag{6.1}$$

where,

Q is the deposited charge (positive or negative) by the particle strike, τ_α is the collection time constant of the p–n junction, τ_β is the ion-track establishment time constant. The time constants τ_α and τ_β are dependent on process technology. The values for τ_α is typically in the range 50–200 ps. On the other hand, τ_β values usually a few picoseconds [6].

In circuit simulations, an independent current source is connected between the drain and body terminals of a MOSFET transistor (see Fig. 6.3).

Depending on logic state, either drain/well and drain/substrate junctions of off CMOS transistors would be vulnerable to a strike. Figure 6.4 shows the equivalent circuit of the inverter given in Fig. 6.3. When input is at logic high, the PMOS transistor would be off and susceptible to a radiation strike at its drain terminal.

In circuit simulations, the effect of a PMOS transistor hit or a "p-hit" would be simulated by a current source taken in upwards direction.

Similarly, for a logic low input, since the NMOS transistor would be susceptible, the current source is drawn in downwards direction to simulate the effect of an NMOS transistor hit or an "n-hit" (see Fig. 6.4).

For terrestrial level applications, the max LET of particles is approximately at 15 MeV cm^2/mg assuming high-energy atmospheric neutrons [6]. A particle with an LET

Fig. 6.3 Circuit level modeling using an independent current source

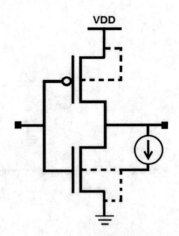

Fig. 6.4 The equivalent circuits for a p-hit (left) and a n-hit (right)

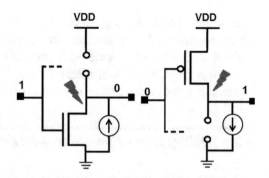

of 1 MeV cm^2/mg deposits around 10 fC/μm along its track; hence an upper bound of 150 fC/μm charge density can be calculated. The collected charge, on the other hand, can be found by multiplying the charge density with the charge collection depth. For newer technologies, the charge collection depth is mostly 1 μm, hence the maximum deposited charge can be assumed as 150 fC at commercial level.

Although the classical double exponential current pulse model in (6.1) is very popular, it can be inaccurate in newer technologies. The affected area in a CMOS integrated circuit due to an ion strike has changed as device feature size has decreased with technology scaling [7]. A single event strike may affect multiple nodes (nearby devices) and nearby well contacts. As a result, a "plateau" in the single event current pulse following the prompt response is observed with higher LET pulses [8, 9].

For deep sub-micron technologies (DSM), the waveform of the corresponding SET produced by the current source does not reproduce that predicted using TCAD (Technology Computer-Aided Design) based device simulations and can lead to different SET amplitudes especially for higher LET (>10 MeV) particles.

The use of ideal double exponential current source alone is not sufficiently accurate, although double exponential current sources still provide a reasonable first-order estimate as a base function model [10].

A mixed-mode simulator may be used to correctly model to model Single Event effects. This simulator combines device level model with standard circuit-level SPICE models and creates a unified simulation environment. This allows selected components in a circuit to be modeled at the device level (i.e. off-biased n-channel transistor in the struck CMOS inverter) while the rest of the circuit is modeled at the circuit level. One advantage would be the direct calculation of voltage and current pulses induced in the struck device by a given particle strike. The limitation of mixed-mode simulator would be the size of the circuit that can be modeled which is usually limited to less than 25 circuit elements.

TCAD based methods also require large computation times although they can achieve a great level of accuracy. It is desirable to model particle strikes as current sources that can be easily injected on circuit nodes for performing quick SPICE simulations.

Researchers have also suggested the use of a combined approach where device simulations are first used to characterize current pulses for ion strikes and then these pulses are later used as inputs to Spice simulations to emulate ion strikes [11, 12].

For this purpose, the data obtained from device simulations are fitted to a double exponential pulse model with appropriate characteristic parameters.

6.3   Mitigation of Single Event Soft Errors

Soft errors created due to propagation of single event transients (SETs) are a significant reliability challenge in modern VLSI. The mitigation of single event transients has taken on growing importance as transistor sizes have decreased. SET mitigation has been a challenging task due to its inherently asynchronous and analog nature [13]. The commercial chip manufacturers need to address the impacts, especially in applications that demand high reliability. These applications are medical systems, military or space applications.

Generally speaking, hardening techniques for soft errors can be classified as system level, device level and circuit level techniques [14]. System-level techniques deal with soft errors at the system architecture level and add redundancy in design to achieve error detection/tolerance ability. For logic circuits, the triple-modular redundancy (TMR) can be implemented at the system board level. In TMR technique, the hardware is copied three times and a majority voting logic is used to suppress any corrupt value from reaching to the output. The system level hardening technique for memory circuits include adding parity bits to the memory word such as hamming codes.

Device-level hardening techniques require fundamental changes to the underlying fabrication technology used to manufacture ICs. The main goal in device-level methods is to mitigate charge collection at the site of particle strike. Silicon on Insulator (SOI) process, for example, limits the charge collection efficiency since the transistor source, body, and drain regions of transistors are naturally insulated from the substrate (see Fig. 6.5). In a SOI transistor, the charge deposition path is limited compared to a bulk device.

Finally, *circuit-level techniques* rely on changes in the circuit design to increase soft error tolerance. In general, circuit level mitigation techniques for single event effects

Fig. 6.5 SOI transistor where body is insulated from the substrate

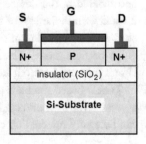

must either filter or dissipate the collected charge or provide some form of redundancy to prevent a soft error.

The easiest circuit level method to reduce the soft error rate of a circuit would be to increase the node capacitance by *adding a capacitor* at that particular node [14]. In its simplest terms, the critical charge of a particular node, can be approximated as $Q_{crit} = C_{node} * V_{DD}/2$, where C_{node} is the node capacitance and V_{DD} is the power supply voltage. Since increasing the node capacitance increases the node critical charge, this fact can be exploited to make a circuit node more robust to radiation. The capacitance method, however, results in large area and power overheads.

Triple modular redundancy is a form of the spatial redundancy techniques. In this method, the logic to be protected is triplicated and a voting circuitry filters the transient. Under normal conditions, all copies of the logic produce the same value. However a particle strike in one of the logic copies can cause a different output value to be produced. The voting circuitry generates a valid result only if at least two calculations agree; hence the incorrect data in one of the outputs will be eliminated. It has been assumed here that more than one strike is less likely and hence is ignored. This method causes large area overhead (>200%) due triplication (see Fig. 6.6).

An alternative to TMR technique would be the *duplication method*. In this case, the combinational logic is duplicated and sent to the asynchronous c-element to filter out the unwanted transient. The c-element (also known as buffer gate) changes its output when both inputs are same. Therefore, a SET transient appearing at one of the inputs will be eliminated (see Fig. 6.7).

Fig. 6.6 Triple modular redundancy

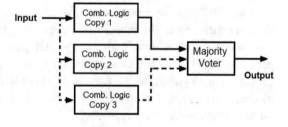

Fig. 6.7 Duplication method along with a c-gate (buffer)

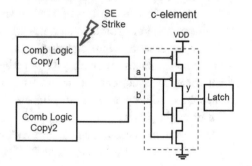

Fig. 6.8 Temporal sampling
latch with variable filtering
delays

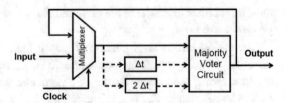

In the case of an unwanted SET pulse, inputs would differ and one of the PMOS or NMOS transistors in either pull-up or pull-down network would be turned off. This would cause output of the buffer gate to float or stay in high impedance state. In other words, the output of the buffer gate would not change and the latch would retain its previous data.

The circuit can be implemented at each latch input to filter out unwanted transients before they get captured. In logic duplication method, the area overhead would be less than the TMR technique; however it remains still high (>100%) due to duplication.

Temporal redundancy techniques separate data signals in time rather than space in order to filter out SETs. These techniques sample the data with different delays and use a majority voter circuit to eliminate the transient [14, 15]. The design shown in Fig. 6.8 shows a hardened latch obtained using temporal redundancy. The circuit includes three separate data paths, a majority voting circuit, and a 2-input multiplexer connected in a feedback loop. In this design, the second and the third data paths incorporate time delays of ΔT and $2\Delta T$, respectively, where ΔT is set equal to the duration of maximum SET width that is targeted to be eliminated.

When the SET noise arrives at the first data path, it is still not available at the outputs of the other paths. Since the majority of the sampling is fed back to the MUX, the transient will be eliminated. After ΔT second, the SET pulse arrives at the output of second data path. The transient will still be eliminated by the voter since the first and the third data paths are unchanged. Although the area overhead is lower compared to spatial redundancy methods, the temporal redundancy technique limits the circuit operating speed due to inherent delay needed for calculation.

Driver Sizing Technique is probably the most popular technique used in soft error mitigation. By increasing sizes of gate's individual transistors, we can increase the output capacitance and transistor drive currents which decrease device sensitivity to radiation [6, 16]. The increase in output capacitance increases the critical charge value (Q_{crit}) of the affected node thereby stabilizing the node against SET effects. At the same time, larger drive strengths or currents of transistors can quickly dissipate the collected charge and reduce the sensitivity to ionizing radiation. Hence, this approach mitigates the SET at the point of origin [14].

As an example, considering the inverter gate shown in Fig. 6.9. The logic 1 value at the input of the first inverter turns off its PMOS transistor making it susceptible to radiation

Fig. 6.9 Driver sizing technique applied to the inverter gate transistors

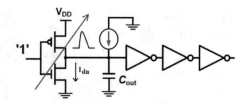

strikes due to reverse-biased substrate to drain p–n junction. The NMOS transistor normally discharges any charge deposited on the output capacitance C_{out}. With driver sizing methodology, the transistor widths are increased such that the SET is dissipated quickly before it is transferred onto the following stage [4]. With sizing, the device current as well as the output capacitance is increased which helps mitigate the impact of an SET. Due to still significant power and area penalties, driver sizing method should be applied selectively. Usually logic stages right before the storage element are hardened using driver sizing since these stages are mostly responsible for single event upsets.

Dynamic Threshold based soft error mitigation scheme uses the increased drive strength of its forward body biased transistors. Normally, the Dynamic Threshold MOS (DTMOS) technique is a power optimization technique that is designed to cut down on the leakage. In a DTMOS logic gate, the body terminal of all transistors is tied to their gates (Fig. 6.10). The high speed operation is obtained by forward bias to switching transistors, while low leakage is obtained by applying zero bias to other transistors. The method requires the use of low power supply that is less than or equal to 0.6 V.

In DTMOS gate, the body-source junction is "forward biased" (at less than 0.6 V) forcing the threshold voltage to drop for the on transistors. Analysis has shown that decreasing threshold voltage increases the critical charge of logic circuits thus providing more immunity to SETs [15]. Hence, this lower threshold effect in DTMOS can be exploited for SET mitigation.

Fig. 6.10 DTMOS logic inverter

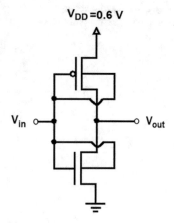

Fig. 6.11 The inverter string
used in HSPICE simulations

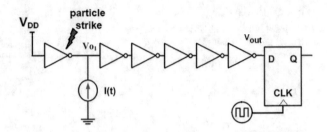

The impact of transistor threshold on critical charge can be examined using a 6-stage inverter string as shown in Fig. 6.11. In our analysis, 65 nm technology with parameters obtained using the Predictive Technology Model in [17] has been considered.

An SE hit was simulated at the output of the first inverter using a double exponential current pulse. In order to examine the impact of reducing threshold on circuit radiation tolerance, the threshold voltage value has been modified via *vth*$_0$ parameter using *delvto* option in HSPICE. The nominal threshold voltage for the NMOS transistor was $V_{th} = 0.29$ V.

Figure 6.12 shows the effect of threshold voltage on the critical charge Q_{crit} for the inverter string. The results show that a 100 mV reduction from normal threshold value increases Q_{crit} at hit node by 33%. This indicates that if threshold voltage can be reduced, the circuit robustness to SETs could be increased. On the other hand, increasing threshold values decrease circuit tolerance to SETs.

Various benchmark circuits were examined in [16] and it was found that the critical charge value obtained for DTMOS configuration to be about 50% more than that of normal body tie configuration. Referring to Fig. 6.10, when the input to the DTMOS inverter is logic 1, the body terminal of an NMOS transistor sees a positive bias in the amount of VDD. Hence, the body terminal gets a forward bias with respect to source

Fig. 6.12 Critical charge
versus device threshold change

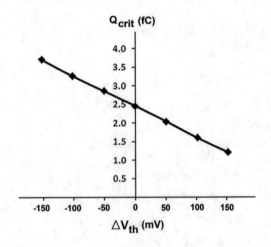

terminal. Since this increases the drivability of the NMOS transistor, a positive SET at the output is easily dissipated via increased current drive.

Therefore, a soft error hardening technique based on the combined use of the standard DTMOS scheme along with driver sizing was proposed [16]. In this case, the combined approach resulted in area savings greater than 30% compared to driver sizing taken alone. This is possible due to the fact that a standard DTMOS gate is more robust compared to a conventional one.

One disadvantage of this technique would be the requirement that separate biasing of transistor body terminals is needed. Hence, either triple-well CMOS or SOI technology is required. It also requires the use of low power supply that is less than 0.6 V.

Researchers have also proposed the *Cascode Voltage Switch Logic (CVSL) gates* in order to increase immunity to soft errors [18, 19].

A CVSL gate has two storage nodes; the function and its complement. In a CVSL gate, NMOS transistors are connected to the inputs. Two complementary NMOS pull-down networks are constructed and then connected to a pair of cross-coupled pull-up PMOS transistors. In Fig. 6.13, the NMOS pull-down configuration on the right is constructed in a similar way to a traditional NAND gate and hence producing the NAND function. The two parallel NMOS transistors, on the other hand, produces the AND function (or the function complement) since it is given inverted inputs.

The fact that the output nodes are connected in a feedback loop by the PMOS transistors results in higher immunity to SET pulses compared to its CMOS counterpart. This feedback connection in fact increases the stability of the logic operation.

CVSL gates may result in increased circuit delays unless the number of inputs is high. This is due to the fact that a CVSL gate in a way acts as a 2-stage logic gate due to feedback connection [18]. The performance of a 4-stage CVSL XOR chain was examined and it was found that the CVSL gates can act 2½ times slower than its CMOS counterpart [19]. The CVSL gate also requires complement of input variables.

In order to protect logic against soft errors with a small area overhead, researchers have also explored *transient filtering techniques* based on the use of pass transistors or the transmission gate (TG) design [20–22]. In the basic TG method, the low pass filtering

Fig. 6.13 CVSL gate structure (NAND gate)

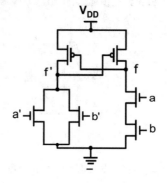

Fig. 6.14 Basic TG gate
method

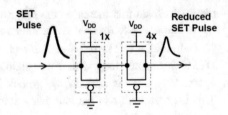

characteristics of TG was utilized in order to reduce the severity of SETs [20]. Here, the devices forming the TG are always made conducting by connecting the gates of PMOS and NMOS pass transistors to *ground* and *power* supply, respectively.

It was claimed by researchers that a two stage TG configuration makes the technique effective if the size of the second stage transistors is made four times larger than the first stage as shown in Fig. 6.14 [20]. In this case, the magnitude of transient pulses is reduced to about 40%. Unfortunately, the method cannot eliminate SET pulses, when the SET pulses with large magnitude occur. It was shown that the magnitude of an SET can be reduced before it is passed on the next stage when TG is used to shield two circuit stages. The TG design can, in the meantime, pass data signals with little change.

Later, a tunable transient filter (TTF) design has been proposed that claimed to have better filtering characteristics [21]. The design consists of two inverters and a number of TGs added in between these inverters as shown in Fig. 6.15. These TGs are also named as the filter gates. It was claimed that the mitigation ability was improved due to the fact that the input signal controlling the filter gates. The hardware overhead of TTF technique, on the other hand, depends on the size of the glitch being eliminated. As the glitch width increase, the number of filter gates increases proportionally.

More recently, a TG scheme with varied gate bias has been proposed which offered superior SET filtering characteristics when compared to basic TG and TTF configurations [22]. This technique relies on applying different gate voltages on the pass transistors of the TG based on the transient duration that needs to be mitigated (see Fig. 6.16). Normally, in order to turn on a TG, an NMOS gate bias (V_{GN}) of V_{DD} and a PMOS gate bias (V_{GP}) of 0 V is needed. However, these transistors can still be turned on with a V_{GN} less than V_{DD} and a V_{GP} greater than 0 V as long as the gate to source-voltage (source-gate) voltage

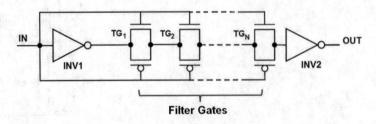

Fig. 6.15 Tunable transient filter (TTF) design

Fig. 6.16 Transmission gate
with varied gate bias

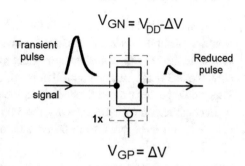

is greater than V_{TN} ($|V_{TP}|$), where V_{TN} and V_{TP} are NMOS and PMOS device threshold voltages, respectively.

The strong filtering effect can be attributed to the increased ON resistance of the pass transistors of TG. This occurs due to reduced gate-source (source-gate) voltages of NMOS (PMOS) pass transistors. This method can mitigate up to 200 ps SET pulse widths using only one stage. This can be achieved by applying different gate voltages on the pass transistors based on the transient width to be eliminated. Table 6.1 shows the gate biases needed to achieve various filtering durations. Last column in this table indicates the signal durations that can be passed without any attenuation.

For larger durations up to 600 ps, the body bias of individual pass transistors can further be controlled which gives additional filtering capability. In this case, the reverse body bias voltages of pass transistors are increased using body terminal of the TG transistors. The increased reverse body bias voltages raise the transistor threshold. This results in increased device ON resistance.

The hardware overhead for basic TG and TTF techniques is mostly linearly dependent on the glitch width. However, for the proposed TG scheme, the hardware overhead remains at $1\times$ for up to 600 ps SET durations.

The delay penalty of proposed TG method is proportional to SET pulse-width that needs to be eliminated. This may not be acceptable for some designs. For faster designs,

Table 6.1 Gate voltages
needed to achieve filtering of
various SET widths

Δ_{SET} (ps)	Gate voltages V_{GN}/V_{GP} (V) needed	Signal duration (ps) Positive/negative pulse
50	0.89/0.36	56/58
75	0.76/0.47	92/95
100	0.71/0.52	128/134
125	0.67/0.55	161/169
150	0.65/0.57	200/208
175	0.64/0.58	232/237
200	0.63/0.59	265/268

the technique can be combined with the driver sizing method in filtering the SETs. In the combine method, driver sizing is used partially to suppress propagated SET at the strike site while the TG suppresses the remaining glitch. In order to keep delay penalty manageable, the TG delay can be controlled using different gate biases for the pass transistors. This would be determined depending on the available time-slack.

In combination with driver sizing, the TG method with varied gate voltage has potential to reduce soft error rate also in faster logic using smaller area overhead.

6.4   Soft Error Mechanisms

There are various mechanisms for radiation induced soft errors in combinatorial logic circuits. SE transients were previously considered as the main cause for combinational logic radiation-induced soft errors until very recently. Other mechanisms such as soft delays, radiation induced clock jitter/race, and SE coupling noise and delay effects should also be considered, especially in applications that demand high reliability. All these mechanisms are gaining importance in newer technologies as technology scaling continues and circuit frequencies increase.

Soft Delay Errors on the rise in newer circuit designs. Soft delay can be described as the amount of delay induced on a CMOS gate due to high-energy particle strike on its sensitive region, which happens only during signal switching [23]. For soft delay to occur, a high-energy particle should strike the sensitive node such as drain node of a CMOS gate's transistor while signal transition is taking place at the output. The current generated due to particle hit can then pull down the signal in the opposite direction causing longer transition time. Incorrect data storage may occur if the delayed signal violates the timing requirements of the storage elements. In other words, a soft delay error (SDE) is created.

As an example, consider the circuit given in Fig. 6.17. A rising input pulse has been connected to the input. While the first inverter is undergoing a falling phase of transition, its PMOS transistor turns OFF and become susceptible to a particle strike. Assuming that a high energetic particle strikes node V_A during this phase of transition, the generated current on PMOS transistor drain due to SE hit (represented by the current source) can pull the signal in positive direction causing longer transition times. The delay effect is observable at the output of the succeeding gate(s), if the path is logically enabled.

The circuit shown in Fig. 6.17 has been constructed in HSPICE using 65 nm technology. Figure 6.18 shows the results for time profile of node voltages V_A and V_C. Two inverters were used after the first inverter to filter out the distortion caused by the SE pulse. All inverters are taken as minimum size inverters. We assume that the node V_A is in rising phase of transition and the particle hit occurs near half the power supply point for maximum delay effect. For a deposited charge of only 15 fC, the soft delay amount (extra delay induced) at the output of the last inverter is 116 ps.

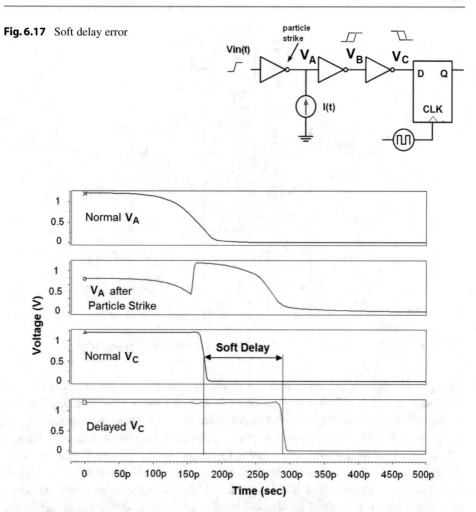

Fig. 6.17 Soft delay error

Fig. 6.18 Waveforms for a 65-nm inverter chain obtained from SPICE simulations

There are three masking effects that can prevent an SET from propagating and being latched by a memory element: Namely, the logical, electrical, and latch-window masking effects. SDEs, on the other hand, can be masked only by logical and latch-window masking. Since there is no electrical masking for soft delay, induced soft delay will always propagate to the next stage and will not get attenuated. Gill et al. have found that a higher number of nodes were sensitive to soft delay effects than SETs due to absence of electrical masking in soft delay propagation [23].

One way to mitigate soft delay error is to use driver sizing methodology, but this happens with the cost of some area and power overheads.

SE-induced clock jitter and false clock pulse events occur when particles inject charge onto clock circuit nodes [24]. SE induced clock jitter occurs when particles strike occurs

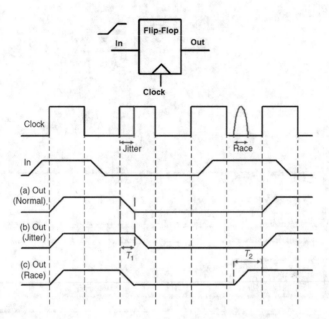

Fig. 6.19 SE-induced crosstalk-jitter and race **a** no clock upset, **b** jitter, and **c** race

during clock edge present. As a result, the clock edge moves back and forth and incorrect data may be stored. Figure 6.19 shows an example of a clock jitter in a flip-flop (FF) configuration. Figure 6.19a shows the output of the FF when there is no clock upset (i.e., in the absence of SE-induced noise), and Fig. 6.19b corresponds to the output of the same FF, with the particle strike occurring during the rising edge of the second clock pulse. In the second case, output signal Out is delayed by T_1 due to radiation induced clock jitter.

Assuming the output signal "Out" is connected later to another storage element, incorrect data storage may occur if the delayed output signal arrives during the set-up time of the receiving sequential.

An energetic particle strike can also create a "false clock pulse" on clock circuit nodes during when there is no clock pulse present. This effect is also known as "radiation induced race" [24]. If the generated pulse carries sufficient magnitude and width, it can be mistaken for a real clock signal. This will cause early latching of data may occur resulting in incorrect data storage.

Figure 6.19c shows the output of flip-flop when a radiation induced race or false clock pulse occurs on the clock node due to particle strike on a clock driver. Due to the formation of this pulse, data is registered early. Hence, the data arrives early by a time T2 as shown. The early latching of data may result in incorrect data storage as well.

For SE clock pulse and jitter effects, hardened pulse generators and pulse latches can be utilized in mitigation.

There are also other noise mechanisms such as *SE crosstalk noise* [25–28] and *SE crosstalk delay* [29–31] effects which occur via interconnect coupling effects. Researchers mostly ignored interconnect coupling in single event analysis of CMOS circuits and focused only on the propagation of SET pulses through logic gates without regards to interconnects between them.

With increasing crosstalk coupling effects, an SET pulse generated on a circuit node is no longer limited to the logic path existing between the hit node and a latch. With interconnect coupling, SE Transients could in fact contaminate electronically unrelated circuit paths which can in turn increase the "Single Event Susceptibility" of CMOS circuits to Single Event Transients.

The technology scaling effects increase *SE crosstalk* effects. Increased cross-coupling effects occur among wires due to reduced distances and increased thickness to width ratio of interconnects. SE Transients on aggressor line can cause larger crosstalk effects than a normal switching induced crosstalk.

Figure 6.20 shows an aggressor-victim pair along with its drivers and receivers. Since inputs of both drivers are held at logic high, the outputs are normally at logic low. An SE hit at the drain of OFF PMOS transistor of the inverter driver causes output to go to logic high (1.2 V) for some pulse duration. The SET voltage created then can affect the victim line through coupling capacitor Cc inducing SE crosstalk noise on the victim. The positive glitch on the victim line can cause false switching of victim receiver giving a logic error.

Hence, SE hits can generate cross-coupling effects that can violate noise margins of gates connected to the affected line and may result in logic errors. Serious effects may occur if the affected line is somewhat important such as a clock line. An upset can be interpreted as a clock edge potentially causing many different bits to be stored incorrectly

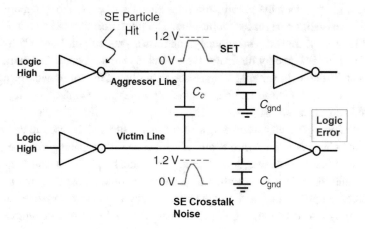

Fig. 6.20 Single event crosstalk noise

causing multiple bit upsets. Due to the large area covered by clocking interconnects there is a higher probability that coupling from neighboring interconnects can have an impact on clocking network [27].

In addition, Triple-Modular Redundancy (TMR) circuits that are used to eliminate Single Event Errors on logic circuits are also susceptible to crosstalk at the inputs of voters [26, 27]. The interconnections between functional units and voters tend to be long and hence sensitive to coupling noise. An undesired coupling among voter inputs can invalidate the data. This is possible since an SET in one redundant circuit may appear on the voting circuit more than once which can cause data corruption.

In order to complement the soft error hardening process, coupling effects need to be considered in the hardening of CMOS logic gates due to technology scaling effects that increase both SE vulnerability and crosstalk effect. Most mitigation methods that apply to regular switching induced crosstalk can be used in *SE crosstalk mitigation*. As an example, *Victim Driver Sizing* methodology is an effective method in reducing SE crosstalk. It reduces both SE crosstalk and normal switching induced crosstalk simultaneously [27]. After sizing up the victim driver, the holding resistance of transistor making the driver reduces while the output capacitance increases. The increased victim output capacitance contributes to stability of the victim line. The increased drive current of victim driver, due to reduced holding resistance, helps mitigate the SE crosstalk noise faster. When victim driver size doubled, the crosstalk noise amplitude reduces by as large as 36% [27].

Another method that is used in regular crosstalk mitigation is the *Aggressor Driver Sizing* methodology. As aggressor driver size is reduced, the amount of induced coupling noise on victim decreases. Aggressor line waveform cannot transition as fast due to its large driver resistance, and as a result, noise amount it will induce on victim line would decrease. This method, however, cannot be used in mitigating SE Crosstalk. Sizing down an aggressor driver reduces the critical charge needed for an SE hit to produce transients; this makes it more vulnerable to radiation transients. As a result, there is more chance to produce SE crosstalk noise on the victim line. The transistors instead should be sized up instead to reduce SE transient magnitude at aggressor gate output. Larger drive strengths of NMOS and PMOS quickly dissipates the collected charge, reduces the vulnerability to Single Event particles. In summary, aggressor sizing has different effects for normal and SE induced crosstalk. Due to conflict in noise optimization, aggressor sizing cannot be used to mitigate SE crosstalk.

If there are no tight area constraints, wire spacing could also be used in controlling the SE crosstalk noise. For a fixed wire width, if we increase wire spacing to its neighbors, its coupling capacitance decreases. Wire spacing is the most effective means for controlling the SE crosstalk but this is with burden of using large routing area and area penalty.

A SE transient pulse generated on the affecting wire (because of particle hits on driver transistors) may also cause increased (or decreased) signal delays on neighbouring (victim) lines via cross-coupling effects if the victim line driver is in switching. In the example shown in Fig. 6.21, a particle hit occurs at the output node of aggressor driver and causes

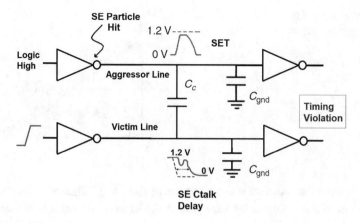

Fig. 6.21 SE crosstalk delay

an SET pulse in the positive direction. The transient then spreads into the victim line via coupling capacitance and becomes present during the switching event. Since the positive crosstalk noise arrives during the waveform is falling, this causes a signal slowdown as shown on the victim line. This effect can be named as the *SE Crosstalk delay*. We may also have SE crosstalk delay effect if victim line waveform is rising but a negative SET transient couples into victim network.

The increase in interconnect delay due to the SE coupling can affect circuit performance because delay changes can violate the setup or hold time requirements of logic storage circuits connected to these receivers.

The amount of induced delay on the victim line is strongly dependent on how the SET waveform on the aggressor line is aligned with respect to the victim waveform. However, circuit designers are often interested in getting the worst delays so large number of nets can be screened quickly before any mitigation is needed. For this, the circuit is first linearized by replacing the drivers by its holding resistances. Once the circuit is linearized, the linear superposition principle is applied [29].

The voltage waveform on the victim line is first obtained assuming there is no perturbance on the aggressor line. Then, the victim net is assumed to be quiet and the SE crosstalk noise (SECN) induced on victim net (due to an SET on aggressor line) is calculated. Finally, the SECN waveform is aligned with the noiseless victim waveform to obtain a composite waveform for worst-case delay calculation (Fig. 6.22) [29, 32, 33].

When calculating for the worst-case delay, a noiseless monotonic victim waveform V_V is assumed and this waveform starts to fall from V_{DD} with a slope of S_V and reaches to zero. One might also consider an SECN signal V_N of arbitrary shape with a maximum height of V_{max} and whose initial and final values are zero, as shown in Fig. 6.22.

Fig. 6.22 Worst-case SE
crosstalk delay calculation
model

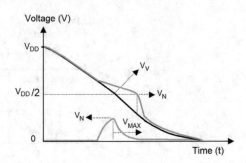

Let us assume the worst-case delay occurs at time t_{wcd}. Then at t_{wcd}, the value of composite waveform obtained by superposing the noise-free victim waveform and the SECN must be equal to $0.5V_{DD}$:

$$V_{DD} - t_{wcd}|S_V| + V_N(t_{wcd}) = V_{DD}/2 \qquad (6.2)$$

Since t_{wcd} corresponds to the largest delay possible, the largest value of V_N should correspond to V_{max}. Then, one obtains:

$$t_{wcd} = \frac{V_{DD}/2 + V_{max}}{|S_V|} \qquad (6.3)$$

In other words, the worst-case victim line delay, t_{wcd} occurs when the noiseless victim waveform ($V_{DD} - t_{wcd}|S_V|$) reaches $0.5V_{DD} - V_{max}$ [29].

The SE Crosstalk Speedup, on the other hand, occurs when a positive/negative SET on an aggressor line couples to the victim line when victim lines is having a rising/falling transition. Figure 6.23 shows an SE particle hitting at the drain of OFF PMOS transistor of the aggressor driver which then causes the output to go towards logic 1 for some pulse duration. The positive SET created, in turn, spreads into the victim line via coupling capacitance and causes a speedup effect on the rising victim line waveform. This may cause victim line signal switch sooner than it normally would resulting in SE crosstalk speedup. The decrease in interconnect delay due to SET coupling can affect circuit performance as the speedup effects may later violate hold time requirements of logic storage circuits connected to these receivers.

Single Event Crosstalk effects can introduce circuit delays that can exceed normal signal switching induced delay after a certain deposited charge level. Although wires might be spaced out enough to prevent timing violations caused by normal aggressor switching during layout optimization, the mitigation taken may not be sufficient to prevent timing errors caused by SET coupling.

In order to complement the SE hardening process, coupling effects among interconnects need to be considered in the SE hardening and analysis of CMOS logic gates due to technology scaling effects that increase both SE vulnerability and crosstalk effects.

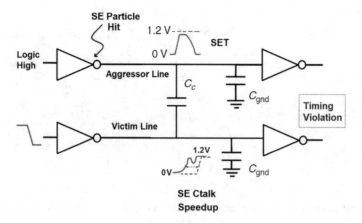

Fig. 6.23 SE crosstalk speed-up

6.5   Single Event Crosstalk Noise Prediction/Modeling

For mission-critical systems, such as military and medical systems, SE coupling effects must be included in the reliability analysis, in addition to SETs. Furthermore, a given network may pass the crosstalk noise check, but it may still pose a threat if SE crosstalk effects are not taken into account properly as discussed previously. It was shown that SE transients can produce crosstalk effects on neighboring wires that can induce logic state changes for lines as short as 100 μm on technologies 90 nm and lower [26].

Traditional SPICE simulators can be used to estimate crosstalk effects in signal lines. While results are accurate, due to millions of nets to be handled, these simulations are time consuming [28]. SE crosstalk modeling and prediction can reduce the computational burden; it would also allow researchers gain insight into the effects of SE pulse transients on the crosstalk noise. Closed-form expressions could be derived and that would allow researchers study crosstalk pulse dependency on various design parameters.

The 4-π circuit template shown in Fig. 6.24 could be used to derive the SE crosstalk model. This template consists of victim and aggressor nets modeled using the 2-π circuits. Here, RC values are calculated based on technology and the geometric information. They are also dependent on where the coupling occurs and the coupling length. Here, R_{1a}/C_{ua} and R_{2a}/C_{da} indicate the upstream and downstream resistance/capacitance for the aggressor line. And, similarly, R_{1v}/C_{uv} and R_{2v}/C_{dv} represent upstream and downstream resistance/capacitance at node 5 of victim net, respectively. Load capacitors, if any, are also lumped in the downstream capacitances.

The current source in Fig. 6.24 represents the particle hit at the output of aggressor driver. The classical double exponential current pulse model is often used in simulations to represent transient currents induced due to radiation which can be inaccurate. The modeling approach used here is based on a combined approach where device simulations

Fig. 6.24 SE crosstalk
modeling template

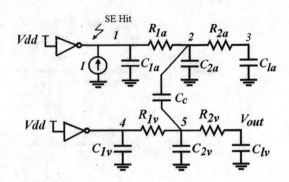

are first used to characterize current pulses for ion strikes and then these pulses are later used as inputs to HSPICE simulations to emulate ion strikes.

For this purpose, the data obtained from device simulations are fitted to a double exponential pulse model with appropriate characteristic parameters. This model assumes that the SE current pulse exhibits an exponential behavior during its rise and decay. The SE crosstalk noise amplitude, on the other hand, is determined mainly by the rising edge of the SET waveform which occurs in picoseconds; hence the effect that this mismatch has during the falling phase has been ignored.

The double exponential model given in (6.4) is composed of an exponential function accounting for the rise in magnitude of the resulting single-event current and another exponential function modeling the decay in magnitude of this current.

$$I(t) = \begin{cases} I_{max}\left(1 - e^{-t/\tau_1}\right) & \text{if } t < t_d \\ I_{max}e^{-(t-t_d)/\tau_2} & \text{if } t \geq t_d \end{cases} \tag{6.4}$$

where,

τ_1 and τ_2 are the rise and the fall time constants of ion-induced current pulse, respectively. I_{max} is its magnitude, and t_d is the delay time for the falling exponential that controls the duration of the plateau effect.

Referring to Fig. 6.24, it has been assumed that both victim and aggressor driver inputs are at logic high and a positive charge deposition occurs on aggressor driver output due to an SE strike. Figure 6.25 shows the new equivalent after drivers represented by the holding resistances of NMOS transistors. Here, the effective resistances R_{da} and R_{dv} model the holding resistances of aggressor and victim drivers, respectively.

The victim driver can be represented approximately by its pull-down network resistance (NMOS transistor holding resistance) since it stays in the linear region. As for the aggressor driver, it can be pre-characterized using stable input (V_{DD}), with the output voltage sweeping from 0 to V_{DD}. In modeling, the medium value of the holding resistance can be selected as an approximation, which is taken at the mid-point of the 0 to V_{DD} range [28].

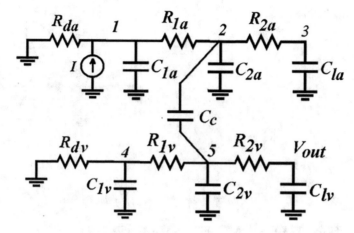

Fig. 6.25 Linearization of driver holding resistances in SE crosstalk modeling

A victim line may be coupled to many passive aggressors. A passive aggressor follows victim waveform and contributes to the stability of the victim line and hence simply cannot be taken as a coupling capacitor at victim coupling point. The equivalent load capacitance at the victim coupling point becomes less than coupling capacitance and can be formulated using coupling/branching admittance concept.

Once, the loading effects of non-switching (passive) aggressors and aggressor tree branches are modeled using equivalent capacitances, the circuit gets reduced to the 4-π template shown in Fig. 6.25.

For calculating the noise at the victim, the aggressor waveform at the coupling node (node 2) is first calculated and then entered to the transfer function between the coupling node and the victim output to obtain victim noise voltage. In order to calculate aggressor coupling node waveform correctly on node 2, the victim loading effect needs to be included. For this, the victim line is reduced into an equivalent capacitor C_{eqv} using the passive aggressor/victim net reduction techniques as outlined in [28]. Figure 6.26 shows the simplified circuit used for calculating aggressor waveform at the coupling node.

The transfer function between the input and coupling node 2 is calculated as:

$$\frac{V_2(s)}{V_{in}(s)} = \frac{1}{st_{a1}+1} \tag{6.5}$$

Fig. 6.26 Decoupled
aggressor line for coupling
node voltage calculation

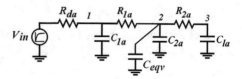

Fig. 6.27 Output voltage calculation

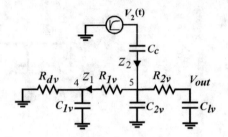

where,

$$t_{a1} = C_{1a}R_{da} + (C_{eqv} + C_{req-1})(R_{da} + R_{1a})$$

C_{req-1} is the equivalent of the π circuit after coupling point.

Note that t_{a1} is the Elmore delay between the input and node 2. Finally, the delayed waveform at coupling node is calculated to be:

$$V_2(t) = I_{\max}R_{da}\left(1 - e^{-t/\tau_{1n}}\right) - I_{\max}R_{da}\left(1 - e^{-(t-t_d)/\tau_{2n}}\right)u(t - t_d) \qquad (6.6)$$

where

$$\tau_{1n} = \tau_1 + t_{a1} \text{ and } \tau_{2n} = \tau_2 + t_{a2}$$

$$t_{a2} = C_{1a}R_{da} + (C_{eqv} + C_{req-2})(R_{da} + R_{1a})$$

In output voltage formulation, the aggressor waveform at coupling location is entered to the transfer function to calculate victim noise as shown in Fig. 6.27.

One obtains:

$$\frac{V_{out}(s)}{V_2(s)} = \frac{s\tau_x}{s\tau_v + 1} \qquad (6.7)$$

where,

$$\tau_v = R_d(C_c + C_{1v} + C_{2v} + C_{lv}) + R_{1v}(C_c + C_{2v} + C_{lv}) + R_{2v}C_{lv}$$

and,

$$\tau_x = (R_d + R_{1v})C_c$$

After inserting (6.6) into (6.7), the victim output noise waveform in time domain is calculated as:

$$V_{out}(t) = I_{max} R_{da} \tau_x \left[\frac{e^{-t/\tau_{1n}} - e^{-t/\tau_v}}{\tau_{1n} - \tau_v} + \frac{e^{-(t-t_d)/\tau_v} - e^{-(t-t_d)/\tau_{2n}}}{\tau_{2n} - \tau_v} * u(t - t_d) \right] \quad (6.8)$$

Noise peak has been traditionally used as a metric to determine if the noise is at an acceptable level. In order to calculate the noise peak, the derivative of (5.28) should be set zero to find the time instant where the maximum occurs. However, the above equation contains three exponential terms and hence it becomes difficult to find a closed-form expression for amplitude. On the other hand, one can obtain a function $f(t)$ which can be used in Newton's iteration method to solve for the t_{peak} where noise peak V_{peak} occurs:

$$f(t) = \frac{dV_{out}}{dt}$$

$$= \frac{I_{max} R_{da} \tau_x}{\tau_v} \left[\frac{\tau_{1n} e^{-t/\tau_v} - \tau_v e^{-t/\tau_{1n}}}{\tau_{1n}^2 - \tau_v \tau_{1n}} + \frac{e^{-(t-t_d)/\tau_v} - e^{-(t-t_d)/\tau_{2n}}}{\tau_{2n}^2 - \tau_v \tau_{2n}} * u(t - t_d) \right] = 0$$

$$(6.9)$$

Then, t_{peak} can be found using:

$$t_{peak_{k+1}} = t_{peak_k} - f(t_{peak_k})/f'(t_{peak_k}) \quad (6.10)$$

This method converges very rapidly after little iteration. The noise peak voltage V_{peak} is found after inserting t_{peak} in (6.8).

A similar victim noise expression can also be obtained for the case that a negative particle strike occurs at the output of aggressor driver. In this case, since both driver inputs are at logic 0, the holding resistances of PMOS transistors should be utilized in noise formulation.

The following steps summarize for the proposed model:

1. Determine the particle LET based on the environment, obtain current pulses from device simulations. Find the rise and the fall time constants τ_1, and τ_2, the delay time t_d, and current magnitude I_{max} of the current source.
2. Calculate the equivalent capacitance value C_{eq-a} for each passive aggressor that is coupled to victim line. This capacitor is then placed in parallel at victim coupling node. Reduce any aggressor tree branches using an equivalent branching capacitance C_{eq-br} at branching point.
3. Repeat the same procedure in step (2) for the victim line, and find an equivalent capacitance value C_{eq-v}. This capacitor updates the value of C_{2a} at the active aggressor coupling point.
4. Calculate the new time constants τ_{1n} and τ_{2n} at aggressor coupling point. Calculate other time constants τ_x and τ_v.
5. Using the iteration formula, find the time that noise peak occurs, which is t_{peak}.

6. In the last step, obtain V_{peak} after inserting t_{peak} in (6.8) and decide whether the noise is acceptable or not.

The model has been tested using over 1000 randomly generated cases to simulate real-time cases in 65 nm technology. The SE crosstalk model has been coded in C++ environment and results are verified by comparing to HSPICE outputs. While the proposed model used a 4-π template, HSPICE simulation utilized a 20-π representation to model the distributed behavior. In Spice modeling, the coupling capacitances were also distributed.

Two parallel interconnects on intermediate layer driven by inverters twitch the minimum size have been assumed. The loads at the end of wires were taken as identically sized inverters. Various interconnect spacing, length and widths were examined. Interconnect lengths were varied from 0.2 up to 2 mm with coupling portion changed and some of the interconnects included some tree branches. The parameter values for these test circuits have been derived using interconnect model given in [17]. The simulated deposited charges, Q_{dep}, were selected in the range 20–150 fC.

Table 6.2 shows the SE Crosstalk noise calculation results for the first 25 cases.

In this table, R_{agg}/C_{agg} and R_{vic}/C_{vic} denote aggressor and victim line resistance/capacitance, respectively. Finally, C_c indicates the coupling capacitance values taken for each case. For the 25 cases considered, the proposed model has an average error of 5.3% when compared to HSPICE. For all cases shown, the percent error stays less than 15%.

Figure 6.28 shows the histogram of SE crosstalk noise calculation error for 1000 different aggressor-victim pair simulated. Considering most of the cases, the percent error stays less than 15%. In Table 6.3, the percentage of nets that fall into the given error ranges has also been given. According to data, about 89% of nets have errors less than 10 percent when predicting the noise peak voltage. On the other hand, around 97% of all nets have errors less than 15%.

According to calculation results, for most of the test cases, the proposed model tends to slightly overestimate the SE crosstalk noise. This might be due to the use of medium value of the holding resistance which might be still overestimating. Another reason for the error might be due the use of dominant pole approximation since the reduction causes some information loss on the transfer function.

For all test cases considered, the proposed model has an average error of 6.16% when compared to HSPICE. The CPU time for these test circuits ranges from 0.025 to 0.30 ms on a 3.0 GHz Pentium IV machine which suggests that the model calculation is at least a 1000× faster than HSPICE.

This chapter discussed soft error generation in combinational logic and circuit level modeling of SE strikes. Various mitigation efforts for soft errors have been discussed with the stress given on the circuit level techniques. Various mechanisms for radiation induced soft errors in combinatorial logic such as soft delay errors, SE-induced clock jitter and race, and finally SE crosstalk noise and delay effects have been discussed in greater detail.

Table 6.2 Experimental results obtained for the first 25 cases

Case #	Q_{dep} (fF)	R_{agg} (Ω)	R_{vic} (Ω)	C_{agg} (fF)	C_{vic} (fF)	C_c (fF)	Noise (mV) HSPICE	Noise (mV) model	|Error %|
1	100	400	500	29	37	50	0.443	0.471	6.32
2	150	350	450	31	40	45	0.670	0.708	5.67
3	75	380	490	30	41	45	0.374	0.407	8.82
4	94	210	225	12	14	27	0.483	0.520	7.66
5	65	225	260	14	17	30	0.368	0.375	1.90
6	90	900	900	64	64	112	0.647	0.667	3.09
7	125	364	357	21	20	48	0.740	0.760	2.70
8	50	350	350	23	23	42	0.380	0.329	13.42
9	120	452	450	27	27	57	0.734	0.746	1.63
10	46	184	232	13	14	23	0.350	0.376	7.43
11	95	356	451	22	27	44	0.533	0.576	8.07
12	39	310	359	30	32	40	0.355	0.361	1.69
13	90	420	455	31	39	52	0.447	0.412	7.83
14	85	290	320	19	23	37	0.410	0.413	0.73
15	83	445	482	23	28	55	0.426	0.446	4.69
16	70	900	900	54	54	113	0.609	0.616	1.15
17	135	421	402	25	24	51	0.859	0.872	1.51
18	90	352	450	25	30	45	0.506	0.520	2.77
19	85	347	413	24	31	43	0.490	0.520	6.12
20	145	690	670	49	43	83	0.792	0.821	3.66
21	25	525	560	35	41	66	0.203	0.180	11.33

(continued)

Table 6.2 (continued)

Case #	Q_{dep} (fF)	R_{agg} (Ω)	R_{vic} (Ω)	C_{agg} (fF)	C_{vic} (fF)	C_c (fF)	Noise (mV) HSPICE	Noise (mV) model	|Error %|
22	35	125	130	11	14	16	0.140	0.155	10.71
23	89	190	210	13	15	23	0.492	0.509	3.46
24	150	460	475	31	40	55	0.710	0.755	6.34
25	65	592	592	37	37	73	0.489	0.508	3.89
Average % error for first 30 cases									5.3%

Fig. 6.28 SECN noise prediction error

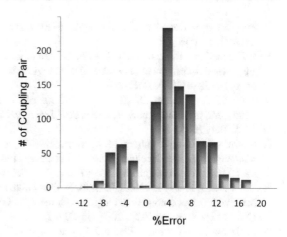

Table 6.3 Percentage of nets that fall into the error ranges

Error range	V_{peak} (%)
Within ±5%	61.70
Within ±10%	88.65
Within ±15%	97.33
Avg. error	6.16

A method to calculate SE crosstalk delay has also been given. The chapter has concluded with SE crosstalk noise prediction where a fast SE crosstalk noise estimation method has been introduced for use in design automation tools.

References

1. P.E. Dodd, L.W. Massengill, "Basic mechanisms and modeling of single-event upset in digital microelectronics," *IEEE Trans. on Nuclear Science*, vol. 50, Issue 3, pp. 583–602, June 2003.
2. R. C. Baumann, "Radiation-induced soft errors in advanced semiconductor technologies," in *IEEE Trans. on Device and Materials Reliability*, vol. 5, no. 3, pp. 305–316, Sept. 2005.
3. T. Heijmen, "Radiation induced soft errors in digital circuits: a literature survey," *Technical report*, Philips Electronics Natl. Lab., Netherlands, August, 2002.
4. B. Jacob, S. W. Ng, and D. T. Wang, *Memory Systems: Cache, DRAM, Disk*, Morgan Kaufmann Publishers, September 2007.
5. R. C. Baumann, "Radiation-induced soft errors in advanced semiconductor technologies," *IEEE Tran. on Device and Mat. Rel.*, vol. 5, no. 3. pp. 305–316, 2005.
6. Q. Zhou and K. Mohanram, "Cost-effective radiation hardening technique for combinational logic," in *Proc. ICCAD*, Nov. 2004, pp. 100–106.
7. S. DasGupta, A. F. Witulski, B. Bhuva, M. Alles, L. W. Massengill, O. A. Amusan, J. R. Ahlbin, R. Schrimpf, and R. Reed, "Effect of Well and Substrate Potential Modulation on Single Event

Pulse Shape in Deep Submicron CMOS," *IEEE Trans. Nucl. Sci.*, vol. 54, no. 6, pt. 1, pp. 2407–2412, Dec. 2007.

8. P. E. Dodd, M. R. Shaneyfelt, J. A. Felix, and J. R. Schwank, Production and propagation of single-event transients in high-speed digital logic ICs, *IEEE Trans. on Nuclear Science*, vol. 51, no. 6, pp. 3278–3284, Dec. 2004.

9. J. Benedetto, P. Eaton, D. Mavis, M. Gadlage, and T. Turflinger, Digital single event transient trends with technology node scaling, *IEEE Trans. on Nuclear Science*, vol. 53, no. 6, pp. 3462–3465, Dec. 2006.

10. S. Kauppila, A. L. Sternberg, M. L. Alles, A. M. Francis, J. Holmes, O. A. Amusan, and L. W. Massengill, "A bias dependent single-event compact model implemented into BSIM4 and a 90 nm CMOS process design kit," *IEEE Trans. Nucl. Sci.*, vol. 56, no. 6, pp. 3152–3157, Dec. 2009.

11. R. Naseer, J. Draper, Y. Boulghassoul, S. DasGupta, and A. Witulski, "Critical charge and set pulse widths for combinational logic in commercial 90nm CMOS technology", *Proc. of the 17th Great Lakes Symp. on VLSI*, 2007, pp. 227–230.

12. S. Uznanski, G. Gasiot, P. Roche, J. L. Autran, and C. Tavernier, "Single event upset and multiple cell upset modeling in commercial bulk 65 nm CMOS SRAMs and flip-flops," *IEEE Trans. on Nuclear Science*, vol. 57, no. 4, pp. 1876–1883, Aug. 2010.

13. V. Ferlet-Cavrois, L. W. Massengill, and P. Gouker, "Single event transients in digital CMOS—a review," *IEEE Trans. on Nuclear Science*, vol. 60, no. 3, pp. 1767–1790, Jun. 2013.

14. R. D. Schrimpf and D. M. Fleetwood, *Radiation effects and soft errors in integrated circuits and electronic devices*, World Scientific, 2004.

15. M. Nicolaidis, "Time redundancy based soft-error tolerance to rescue nanometer technologies," in *Proc. of the 17th IEEE VLSI Test Symp.*, 1999, p. 86.

16. S. Sayil and N. B. Patel, "Soft Error and Soft Delay Mitigation Using Dynamic Threshold Technique," *IEEE Trans. on Nuclear Science*, vol. 57, no. 6, pp. 3553–3559, Dec. 2010.

17. Predictive Technology Model (PTM), http://www.eas.asu.edu/~ptm, 2019.

18. M. C. Casey, et al., "HBD using cascode-voltage switch logic gates for SET tolerant digital designs," *IEEE Trans. on Nuclear Science*, vol. 52, no. 6, pp. 2510–2515, Dec. 2005.

19. H. Hatano, "Single event effects on CVSL and CMOS exclusive-OR (EX-OR) circuits", *RADECS*, 2009, pp. 133–137.

20. J. Kumar, M.B. Tahoori, "Use of pass-transistor logic to minimize the impact of soft errors in combinational circuits", in WKSP on SELSE, pp. 67–74, 2005.

21. Q. Zhou, M. R. Choudhury, and K. Mohanram, "Tunable transient filters for soft error rate reduction in combinational circuits," in *Proc. Eur. Test Symp.*, Verbania, May 2008, pp. 179–184.

22. S. Sayil, A. H. Shah, Md. A. Zaman, Md. A., Islam, "Soft error mitigation using transmission gate with varying gate and body bias", *IEEE Design & Test*, vol. 34, no. 1, pp. 47–56, 2017.

23. B. S. Gill, C. Papachristou, and F. G. Wolff, "Soft delay error effects in CMOS combinational circuits," in Proc. 22nd VLSI Test Symp., 2004, pp. 325–330.

24. N. Seifert, P. Shipley, M.D. Pant, V. Ambrose, B. Gill, "Radiation-induced clock jitter and race," *Proc. Int. Phys. Reliab. Symp.*, pp. 215–222, 2005.

25. A. Balasubramanian, Sternberg, B. L. Bhuva, L. W. Massengill, "Crosstalk effects caused by single event hits in deep sub-micron CMOS technologies," *IEEE Trans. on Nuclear Science*, vol. 53, no. 6, Dec. 2006.

26. A. Balasubramanian, O. A. Amusan, B. L. Bhuva, R. A. Reed, A. L Sternberg, L. Andrew, L. W. Massengill, D. McMorrow, A. Sarah, J.S. Melinger, "Measurement and analysis of interconnect crosstalk due to single events in a 90 nm CMOS technology," *IEEE Trans. on Nuclear Science*, vol. 55, no. 4, pp. 2079–2084, 2008.

27. S. Sayil, A. B. Akkur, N. Gaspard, "Single Event Crosstalk Shielding for CMOS Logic", *Microelectronics Journal*, vol. 40, no. 6, pp. 1000–1006, 2009.

28. S. Sayil, V. K. Boorla, S. R Yeddula, "Modeling Single Event Crosstalk in Nanometer Technologies", *IEEE Trans. on Nuclear Science*, vol. 58, no. 5, pp. 2493–2502, 2011.
29. S. Sayil, Y. Yao, "Single Event Coupling Delay Estimation in Nanometer Technologies", *Analog Integrated Circuits and Signal Processing*, vol. 86, no. 2, pp. 215–225, February 2016.
30. S. Sayil, Li Yuan, "Modeling Single Event Crosstalk Speedup in Nanometer Technologies", *Microelectronics Journal*, Volume 46, Issue 5, pp. 343–35, May 2015.
31. S. Sayil, J. Wang, "Coupling induced soft error mechanisms in nanoscale CMOS technologies", *Analog Integrated Circuits and Signal Processing*, Volume 79, Issue 1, pp 115–126, April 2014.
32. V. Rajappan, S. S. Sapatnekar, "An efficient algorithm for calculating the worst-case delay due to crosstalk", *Proc. 21st Int. Conf. on Comp. Design*, 2003, pp. 76–81.
33. F. Dartu, L. T. Pileggi, "Calculating worst-case gate delays due to dominant capacitance coupling." *Proc. ACM/IEEE Design Automation Conf.*, 1997, pp. 46–51.

Thermal Noise

7.1   Introduction

With the advancement of very large scale integration (VLSI) design and technology, thermal gradients across a chip has become a major factor in determining the chip performance and circuit reliability. With the integration of billions of transistors on a single chip, the total power consumption and power per unit area is continuously rising. As power consumption is usually dissipated as heat, the chip operating temperature increases. In high-performance chips, the peak chip temperature can rise up to 160 °C in 65-nm technology node and is expected to rise even more [1]. The high temperature adversely impacts circuit performance, reliability and robustness. Moreover, the employment of different power management techniques and diverse activities causes the chip to experience non-uniform temperatures across chip area. On-chip temperature gradient can be as high as 50 °C [2]. This can cause on chip temperature gradient to be real large. In addition, for some applications such as military, the circuits needs to function over a wide range of temperatures ranging from −55 to 125 °C [3, 4].

If heat is not removed at a rate equal to or greater than its rate of generation, junction temperatures will rise and higher junction temperatures reduce mean time to failure (MTTF) for the devices. Device reliability has a direct impact on the overall system reliability; therefore impact on temperature is an important issue.

The temperature dependency of a MOS transistor current can usually be explained using the alpha power model for short channel device. The alpha-power law is commonly used to model the propagation delay T_p of a CMOS device. This relation can be given as [5]:

$$T_P \alpha \frac{C_L V_{DD}}{I_D} = \frac{C_L V_{DD}}{\mu(T)(V_{DD} - V_{TH}(T))^{\alpha}} \tag{7.1}$$

S. Sayil, *Noise Contamination in Nanoscale VLSI Circuits*, Synthesis Lectures on Digital Circuits & Systems, https://doi.org/10.1007/978-3-031-12751-9_7

where,

C_L is the load capacitance; I_D is drain current and α is the positive technology constant. $V_{th}(T)$ and $\mu(T)$ are the temperature dependent threshold voltage and carrier mobility, respectively.

Here, both the threshold voltage and carrier mobility decrease with increasing temperature. However, Eq. (7.1) shows that they have conflicting effects on the drain current.

The impact of temperature on buffers is not straightforward as two important parameters, carrier mobility and threshold voltage, has conflicting effect on device current when temperature changes [5]. As temperature increases, the carrier mobility decreases which causes the device current to reduce. On the other hand, increased temperature decreases device threshold voltage causing drive current to increase. Depending on the supply voltage, device current may increase or decrease. At nominal supply voltage (~1 V), change in carrier mobility dominates over the change of threshold voltage which causes the device current to decrease with increased temperature. However, when operating at relatively low bias voltages this effect is reversed; the effect of the threshold-voltage variation with temperature becomes the dominant factor. In other words, the device current increases with elevated temperatures.

There exists a supply voltage such that the two effects compensate each other making the current insensitive to temperature variation. This point is known as the Zero Temperature Coefficient (ZTC) Voltage or Temperature insensitive voltage V_{INS} [4]. As technology scales this voltage approaches nominal voltage.

The impact of temperature on interconnect, on the other hand is relatively straightforward. First of all, the effect of temperature can safely be ignored for inductance, and the effect on capacitance is also very low. The most important temperature affected parameter is the interconnect resistance, R. Concerning metal wires, an increase in temperature causes a linear increase of the metal resistance. The effect can be expressed as the following:

$$R(T) = R_0[1 + \alpha_R(T - T_0)] \tag{7.2}$$

where,

R_0 is the resistance of interconnect at a nominal temperature, α_R is the temperature coefficient of resistance and T represents the local temperature of the wire.

The amount of change in resistance can be very large. Over the military-specified temperature range, Cu wire resistance can change up to 72.2% with temperature [2].

7.2 Compensation of Thermally Induced Clock Skew

At nominal supply voltages (~1 V), $V_{DD} - V_{TH}$ difference becomes less sensitive to thermally-induced threshold variation, and hence mobility effect determines the temperature impact on the drain current. As a result, device current decreases with increased temperature at nominal supply voltage. When operating at relatively low bias voltages, however, this effect is reversed. In other words, the device current increases at high temperature due to the fact that threshold voltage effect dominates.

Thus, there are two types of regions and in between these two regions there must have an intersecting point where drain current will neither increases nor decreases. This happens for a specific supply voltage which is called temperature insensitive point (V_{INS}) or a zero temperature coefficient (ZTC) point. At V_{INS}, the two effects basically compensate each other making the current insensitive to temperature variation [2, 4].

Researchers have proposed various schemes in order to maintain driver strength constant with respect to temperature effects [2, 4, 6–8] so that delay uncertainty or clock skew can be avoided.

Temperature insensitivity on the drain current can basically achieved by simply controlling the gate-source voltage via power supply scaling [6, 7]. In order to examine the impact of temperature on drain current, we have performed HSpice simulations using 45 nm Predictive technology model [9]. Figure 7.1 plots the drain saturation current of NMOS and PMOS transistors as a function of supply voltage (V_{DD}) at 25 and 125 °C. The inset of Fig. 7.1 shows the diode connected NMOS and PMOS transistors where V_{DD} voltage is varied from 0 to 1 V. From Fig. 7.1, we can determine the value of V_{DD} (or V_{GS}) at which the drain current becomes insensitive to temperature or the value of V_{INS}. Due to differences in electron and hole mobility, PMOS and NMOS devices have different values of V_{INS}. Our simulation results show that the V_{INS} value is around 0.685 V for PMOS and about 0.520 V for NMOS device, respectively.

The disadvantages are that the values found for V_{INS} are lower than the nominal V_{DD}, hence they result in large circuit performance penalties due to reduced drain current. Moreover, the PMOS and NMOS transistors have different values of V_{INS} limiting the effectiveness of this technique for thermal compensation.

Several other schemes have also been proposed to maintain the driver strength constant using V_{INS}. The work in [6] has proposed a design methodology based on optimizing the supply voltage to obtain a temperature insensitive operation.

In [4], authors have proposed a dynamically adaptive clock buffer that used both the buffer current control and body-bias control techniques. In buffer control technique additional pull-up and pull-down branches are created with these branches providing additional amount of current at elevated temperatures to compensate for the reduced buffer current. Furthermore, the body bias of all transistors is also adjusted to reduce the threshold and boost the current. Since reduced threshold voltages increase device speed, this fact is also used to compensate for the speed degradation at high temperatures. While this

Fig. 7.1 Drain saturation current as a function of supply voltage and temperature: **a** NMOS transistor, **b** PMOS transistor

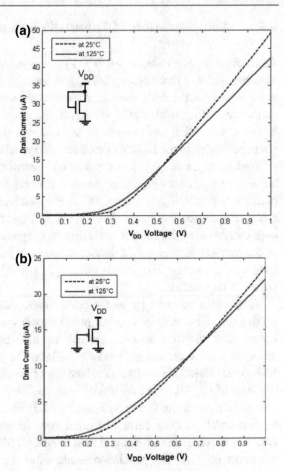

technique achieves good temperature insensitivity, designing the temperature sensor and wave-shaping circuit can be complex and area costly and may increase power dissipation greatly.

In multi-threshold technique proposed in [8], the temperature insensitivity was obtained by proper interleaving low threshold (*lvt*), standard threshold (*stv*) and high threshold voltage (*hvt*) devices. The temperature insensitivity, however, was obtained only for supply voltages between 0.6 and 0.8 V.

Wolpert and Ampadu [2] proposed the use of programmable temperature compensation devices (PTCDs) to maintain buffer strength constant with respect to temperature. In this technique, a series PMOS and an NMOS transistor (PTCD) are placed on the pull-up and pull-down networks, respectively.

Unlike previous methods which manipulate V_{INS} by controlling the power supply, the PTCD method achieves temperature insensitivity by controlling the gate bias of series

Fig. 7.2 The use of PTCD
devices in a CMOS gate

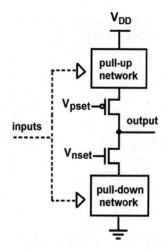

inserted transistors. These series transistors (PTCDs) are placed below and above the pull-up and pull-down networks, respectively as can be seen in Fig. 7.2. They are basically placed near the output to prevent a virtual power supply.

Here, the gate control voltages of PTCDs (V_{pset} and V_{nset}) are set at the temperature insensitive points. By setting V_{pset} and V_{nset} at corresponding V_{INS} values, we can achieve temperature insensitivity. The serially inserted PCTDs basically set the current around the insensitive point. Compared to power supply scaling method, the PTCD method has the advantage that it allows the use of nominal supply voltages. It also does not require a temperature sensor. On the other hand, the current level set is less than the full drive current, the methodology results in large performance overhead.

In order to illustrate this method, we have used the structures shown in Fig. 7.3. These structures would behave similar to individual transistors shown in Fig. 7.1 but with the added benefit that the gate overdrive voltages can set the biasing current. For each structure, Figs. 7.4 and 7.5 show the scaled drain currents (current at specified temperature scaled by the current at −55 °C) as a function of temperature.

Results show that the drain current increases with increasing temperature when control voltage (V_{nset}/V_{pset}) is set at a smaller value than V_{INS}. By controlling the gate voltages of these series transistors and setting them at V_{INS}, the currents can be set such that temperature insensitivity is obtained.

The serially inserted PCTDs basically set the current around the insensitive point. The middle curve in Figs. 7.4 and 7.5 indicates that the current behaves pretty much independent of temperature. However, since the current level set is less than the full drive current, the methodology, however, results in large performance overhead.

Fig. 7.3 Controlling the gate voltages of PTCDs

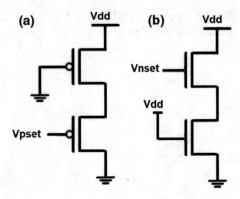

Fig. 7.4 Adjusting V_{nset} for temperature insensitivity

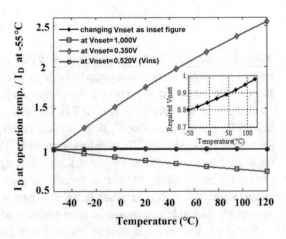

Fig. 7.5 Adjusting V_{pset} for temperature insensitivity

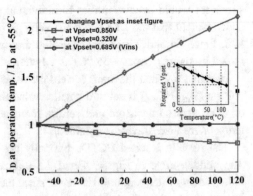

7.3   Thermally Induced Soft Errors

As supply voltages and node capacitances are reduced, circuits become much more susceptible to radiation due to reduced critical charge of circuit nodes. Moreover temperature
may further increase circuit sensitivity to radiation which can in turn increase the soft error
rate. Increased temperature affects the driving strength of transistors by increasing driver
resistance. Therefore, if thermal impact is not considered properly, standard measures
such as driver sizing may fail as temperature increases, causing new reliability issues.

Single-event error rates are a strong function of SET pulse widths, and parameters that
control SET pulse widths (drift, diffusion, bipolar effects, etc.) are strong functions of
operating temperature. As space-borne electronic circuits are often required to operate in
extreme temperature environments, effects of temperature changes on the soft-error rate
are of vital importance [10, 11].

The total charge collected at struck node is the sum of drift, diffusion, and parasitic
bipolar amplification components [12]. Drift collection is dependent on carrier veloc-
ity which is a function of mobility and electric field strength, whereas carrier mobility
decreases with increasing temperature, resulting in less charge collected by drift. But the
decrease in mobility is not significant as the funnel-assisted drift process last only about
10–20 ps in a typical technology, which is much shorter, compared to the entire digital
SET pulse width [13]. The amount of charge collected by diffusion process depends on
the minority carrier diffusion length, which itself depends on the am-bipolar diffusion
coefficient and minority carrier lifetime. The diffusion constant increases and the lifetime
decreases with increasing temperature and the combined result is a diffusion charge that
is only weakly influenced by temperature [14].

When a particle strike to the drain/well junction, numerous electron–hole pairs generate
along the ion track and electrons can be collected quickly. Holes can only escape through
the body tie or body contact. Residual holes left in the well increase the well potential and
lower the source/well potential barrier. As a result source injects electrons into the channel
and these electrons can be collected at the drain where they add to the original particle-
induced current. This can cause an increased charge collection which is known as bipolar
transistor effect, where the source acts as the emitter, the channel as the base region, and
the drain as the collector [15]. In case of bulk technologies, as temperature increases, the
resistance between the body contact and the bulk under the gate increases, resulting in a
higher well potential and corresponding larger bipolar amplification. On the other hand the
beta of the parasitic transistor also increases with increasing temperature. The increment
of the beta rather than the resistance is the dominant factor for the enhancement of bipolar
amplification with rising temperature. Therefore bipolar amplification is the dominating
factor for temperature dependence of digital SET pulse width [16].

The impact of temperature on SET can be illustrated by a 6-stage inverter chain using
45 nm technology as shown in Fig. 7.6. In the inverter chain the SET current pulse of 50
fC is applied at the output of driver inverter (first inverter) and the input is connected to

Fig. 7.6 Thermal impact on SET waveform **a** 6-stage inverter chain, **b** output waveforms

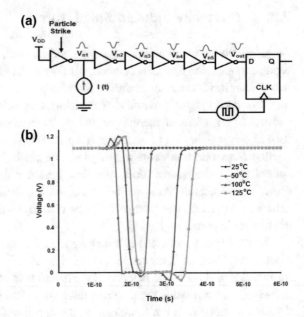

the V_{DD} (logic 1). The output is observed in four temperature conditions of 25, 50, 100 and 125 °C with the same driver size of 16.45× (740 nm). It was observed that although SET pulse was mitigated at 25 °C, but there was a significant amount of SET pulse output at temperatures 50, 100 and 125 °C.

When radiation strike at CMOS circuit node, increase in temperature might change the SET noise to exceed the noise margin of subsequent gates causing a wrong data value to propagate. Hence, mitigation efforts should be done with the thermal impact kept in mind, otherwise designs might fail.

7.4   Analysis and Mitigation of Thermally Induced Single Event Crosstalk

This section reports that varying temperature profiles on neighboring interconnects can further alleviate SE crosstalk noise (SECN) effects. The increased temperature affects both interconnect resistance and driving strength of transistors, hence crosstalk noise will be impacted. If temperature induced effects are not properly considered, standard measures taken for mitigation such as wire spacing, gate sizing etc. may fail causing design failure. We propose a mitigation method for thermally induced SE crosstalk based on adaptive body biasing of devices and a temperature sensor. Simulation results demonstrate that, the proposed method can mitigate thermally induced crosstalk noise by 86% on average.

Previous work in [17] has shown that radiation transients can induce cross coupling noise effects on neighboring lines that can induce logic state changes for interconnects as small as 100 μm on technologies 65 nm and below. In some cases, the induced noise amplitude can be even larger than a normal switching induced crosstalk.

The increased temperature affects both wire resistance and driving strength of driver gate transistors. The resistance of an interconnect increases linearly with temperature due to positive temperature co-efficient of materials used such as copper. The impact of temperature on a driver is not clearly defined due to the fact that carrier mobility and device threshold voltage has conflicting effects on the current when temperature changes [18]. As temperature increases, the carrier mobility decreases which causes the device current to reduce. With increased temperature, device threshold on the other hand reduces causing drive current to increase. Depending on the supply voltage level, device current may increase or decrease. At nominal supply voltage (~1 V), change in carrier mobility dominates over the change of threshold voltage which causes the device current to decrease with increased temperature.

Hence, if a victim line is at a higher temperature, we can expect more SECN on victim line at nominal supply voltage due to reduced driver strength and increased interconnect resistance of the victim line. With thermal contribution, total SECN amount may exceed the noise margin of the subsequent gate causing a wrong value to be propagated. In other words, although a given net may pass the normal crosstalk noise check, it may still pose a threat if thermal effects are not properly considered.

With increased coupling effects, the Single Event Transient (SET) generated on a circuit node may affect multiple logic paths due to strong coupling among wires. Figure 7.7 shows aggressor-victim pair along with its drivers and receivers. Since inputs of both drivers are held at logic 1, the outputs are normally at logic 0. An SE hit at the drain of OFF PMOS transistor of the inverter driver causes the output to go towards logic 1 for some pulse duration. The SET voltage created, in turn, can affect the victim line through coupling capacitor Cc inducing SE crosstalk noise on the victim.

In Fig. 7.7, the solid line (blue waveform) indicates the induced SECN on victim line at low temperature. Now, assume that the victim line experiences a higher temperature due to thermal variation on the chip. In this case, the reduced victim driver strength and

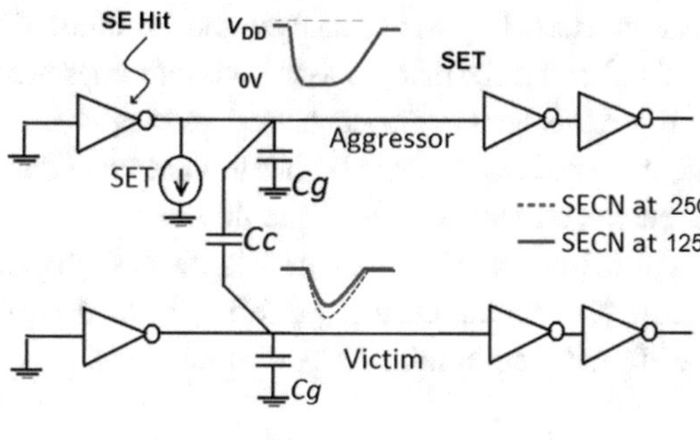

Fig. 7.7 SECN induced on the victim net at different thermal conditions

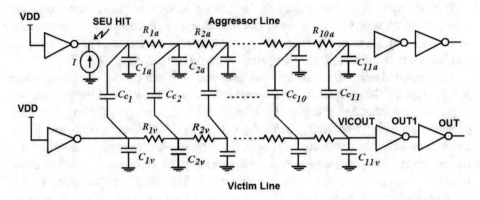

Fig. 7.8 Simulation setup used for SECN analysis

increased victim line resistance can cause SECN get stronger as shown by the dashed line (red waveform). When victim driver experiences high temperature, it loses its capacity to dump the excess charge that is generated by the negative SET waveform on aggressor line. Consequently, the same deposited charge induces more negative crosstalk glitch on the victim receiver. With thermal contribution, total SECN may exceed the noise margin of the subsequent gate causing an incorrect value to be propagated. In other words, although a given net may pass the normal crosstalk noise check, it may still pose a threat if thermal effects are not properly considered.

In order to study the impact of temperature on SECN, consider the simulation set-up shown in Fig. 7.8. In this setup, a 10-π model with distributed coupling capacitances is used for every 0.1 mm of wire to represent the wire distributed behavior. In the modeling, the inductance effects have been ignored and capacitive coupling is assumed as the dominant mechanism for crosstalk [19].

Two parallel wires on intermediate layer in 45 nm technology have considered with wire dimensions taken as follows: the width (W), spacing (S) is 0.14 μm, and wire thickness T is 0.35 μm [9]. A wire length of 0.5 μm is taken for the intermediate wire. It is assumed that both aggressor and victim driver sizes are 3 times the minimum size (e.g. $W_p/W_n = 270$ nm/135 nm) and the loads at the end of the wires are minimum sized identical inverters. In order to analyze the SE induced crosstalk noise, we assumed a positive SET pulse and for that we assumed both lines are at "logic 0". Hence, the aggressor and victim driver inputs are connected to "logic 1". In this case, normally aggressor driver output would be at "logic 0" but it would be taken to "logic 1" if there is a sufficient SE hit charge on output node of the driver.

An SE hit of 75 fC charge has been simulated at the output of aggressor driver. A double exponential current pulse has been used in our analysis as an approximation to waveforms seen in mixed-mode simulations.

Initially, both the aggressor and victim lines are assumed to be at room temperature (at 25 °C) and victim line end waveforms have been recorded. Simulation has been repeated for the case that victim line is at a high temperature (125 °C) and the aggressor is at room temperature. Figure 7.9 shows the victim end waveforms for the same temperature profile (25 °C) on aggressor and victim lines. Referring to Fig. 7.9, the SECN glitch induced on the victim line is less than the noise margin of the victim receiver, hence no error is observed. However, when the victim net temperature becomes higher (125 °C), a significant glitch above the noise margin of the receiver has been observed. As a result, an error occurs at the output of last receiver gate as shown in Fig. 7.10.

This indicates that although SECN can be mitigated at normal temperature, high temperature can render the protection ineffective. This reduced mitigation occurs due to increased victim line resistance and weak victim driver.

Figure 7.11 shows the relation between the temperature and the thermally induced SECN on victim line. In this simulation, deposited charge amounts of 75 and 125 fC have been assumed along with randomly selected victim driver sizes. In order to obtain the temperature induced SECN, the amplitudes of the SECN on the victim line were recorded at room and high temperature, with the difference being recorded as the temperature induced SECN. The temperature-induced SECN grows in proportion to the temperature

Fig. 7.9 Victim end waveforms for normal condition (aggressor and victim line temperatures at 25 °C)

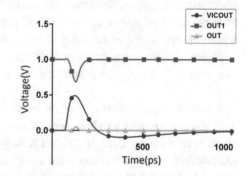

Fig. 7.10 Victim end waveforms for high temperature condition (at 125 °C)

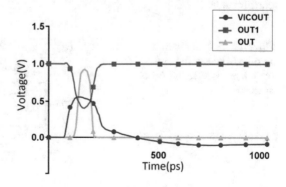

Fig. 7.11 Temperature
induced SECN contribution

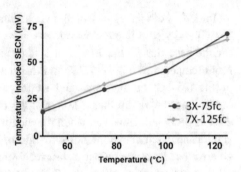

of the victim line and the driver, according to the data. This extra noise amount can be
sufficient to switch a victim driver.

When the victim line is at a high temperature, due to thermal effects, the total SECN
may exceed the noise margin of the subsequent gate causing a wrong value to be propagated. Hence, previous mitigation measures such as driver sizing taken may fail. This
happens due to reduced driver strength and increased interconnect resistance of the victim
line.

One solution might to be to increase driver size to absorb temperature induced errors
which would in turn increase the circuit area. The victim line delay would still vary based
on temperature effects in this solution. If the logic path that includes the victim gate is
among the critical paths, the delay changes may not be well tolerated.

Hence, a technique to compensate for the thermally induced SECN effects by controlling body bias of active transistors and using a temperature sensor has been proposed in
[19].

Normally, each transistor forming a gate has a *zero-body bias*, meaning PMOS and
NMOS transistor body terminals are tied to supply and ground respectively. This is usually
called as *zero-body bias*. In the "Forward Body Biasing" (FBB) scheme shown in Fig. 7.12
[20], the PMOS transistor body terminal is tied to a voltage lower than V_{DD} (V_{BP}) and the
NMOS body terminal is tied to a voltage higher than ground (V_{BN}) to enhance transistor
performance.

Fig. 7.12 A driver with **a**
zero-body bias, **b** forward
body-bias via VBP and VBN
as shown

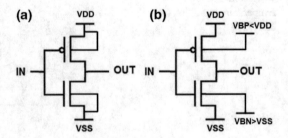

The body-bias can alter the threshold voltage of a device exploiting the body effect of MOS transistors. The body effect can be modeled by:

$$V_{TH} = V_{T0} + \gamma \left(\sqrt{|V_{SB} + 2\phi_f|} - \sqrt{2\phi_f} \right) \tag{7.3}$$

where,

V_{T0} is the threshold voltage under zero substrate bias,

γ is the body effect parameter,

V_{SB} is the voltage between the source and bulk terminals,

$2\phi_f$ is the surface potential.

With FBB, transistor threshold voltages reduce which increase transistors currents and circuit speed.

In the proposed methodology, zero body-bias is applied to transistors at room temperature and a forward body-bias is applied at higher temperature. For an NMOS transistor, if body voltage V_{BN} increases gradually with the increase of temperature, it can be made insensitive to temperature. Similarly, the body voltage V_{BP} needs to be swept down from V_{DD} gradually with the rise of temperature for the PMOS device.

The required body biases are provided by the temperature sensor design shown in Fig. 7.13. Temperature dependency is obtained at gate-source voltage of a diode-connected MOS transistor when biased with a constant current [21]. Here, the reference current is formed by resistor R1 and the transistor, M1 and the gate to source voltage of M1 is applied to M2. The voltage generated across resistor R2 is at 0.878 V (@ 25 °C) and increases at 1.07 mV/°C.

Further amplification is made with M3 and M4. While the input voltage of M3 increases with temperature in proportion, due to inverting characteristics of the inverter, output will decrease accordingly. The resulting voltage is applied to the body of PMOS transistor (V_{BP}). This voltage value is set to 0.93 V at 25 °C, and it reduces with the increase of temperature. The voltage decreases at a sensitivity rate of 3.61 mV/°C. On the other than, the voltage drop at the load resistor R4 is fed to the body of NMOS transistor (V_{BN}). At 25 °C, V_{BN} is at 5 mV and it reaches to 530 mV at 125 °C. All values can be

Fig. 7.13 The temperature sensor design

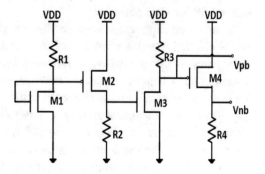

Fig. 7.14 The voltages
generated by the temperature
sensor

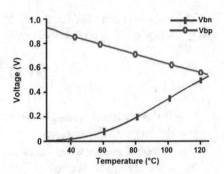

customized according to the design criteria. The V_{BP} and V_{BN} values obtained from the
sensor are shown in Fig. 7.14.

The proposed methodology has been tested using 1000 randomly generated cases to
simulate real-time cases in 45 nm technology. Various driver sizes, interconnect length,
widths and spacing were examined. Interconnect lengths were varied between 0.5 and
2 mm with coupling portion amount changed. The parameter values for these test circuits
have been derived using interconnect model given in [9]. The deposited charge, Q_{dep}, has
been changed from 25 to 150 fC.

Table 7.1 shows the results obtained for the first 15 cases. Referring to these cases, the
thermally induced SECN amount has been reduced by up to 90% without requiring any
change on driver size. Another advantage of this technique is that the delay uncertainty
has been greatly eliminated. Referring to the cases shown in Table 7.1, the delay penalty
due to high temperature on victim line increases the victim line delay by 26% on average.
After using mitigation, the delay increases only up to 7.7%.

For all cases considered, the results indicate that the technique reduces thermally
induced SECN contribution by 86% on average. For all randomly generated cases, the
average delay overhead remains low at 8%. Considering the average delay penalty of
30% paid for all cases, this technique is very advantageous for keeping the delay under
control for critical paths. The proposed sensor generates up to 10% error in generating the
needed body voltages. Hence, further calibration may be needed to eliminate all thermally
induced SE transient pulse.

In summary, high temperature profiles on victim interconnects can further increase
SECN effects impacting reliability of mitigated circuits. A mitigation method for thermally induced SE crosstalk was proposed based on adaptive body biasing of driver
transistors and a temperature sensor. In the proposed method, a temperature sensor senses
the temperature profile and provides the required body-bias voltage for victim drivers to
compensate for the temperature effects. Simulation results demonstrate that, the proposed
method can mitigate thermally induced SECN up to 86% and keep delay penalty within
8% range. The technique is advantageous if victim gates lay among a critical path since
it greatly reduces the temperature induced path delay which can be really large.

Table 7.1 The simulation results for the first 15-cases

Case #	Dep. charge (fC)	SECN amplitude (mV)			Thermally induced SECN reduction (%)	Victim line delay (in ps)			Delay overhead (%) at 125 °C	
		25 °C	125 °C	After mitig. (125 °C)		25 °C	125 °C	After mitig. (125 °C)	Before mitig.	After mitig.
1	77	483	538	493	81.82	72	87	77	20.64	6.73
2	57	468	527	474	89.83	109	131	116	19.97	5.75
3	35	387	441	394	87.04	239	287	251	20.10	5.01
4	45	477	508	480	90.32	333	435	364	30.54	9.07
5	60	485	513	489	85.71	663	871	729	31.39	9.93
6	73	492	521	495	89.66	992	1313	1021	32.30	2.89
7	90	516	564	525	81.25	144	179	160	24.56	11.01
8	105	553	598	561	82.22	218	282	243	29.03	11.36
9	72	485	539	493	85.19	218	267	237	22.46	8.41
10	83	461	515	470	83.33	328	411	365	25.12	11.27
11	48	483	512	490	75.86	476	576	505	21.09	6.09
12	59	485	513	491	78.57	711	869	762	22.26	7.24
13	56	399	439	403	90	363	478	397	31.71	9.38
14	71	448	514	458	84.85	723	957	776	32.43	7.32
15	82	499	543	508	79.55	1083	1443	1122	33.27	3.57

References

1. S. Im and K. Banerjee, "Full chip thermal analysis of planar (2-D) and vertically integrated (3-D) high-performance ICs," in *Proc. Int. Electron Device Meeting*, 2000, pp. 727–730.

2. D. Wolpert and P. Ampadu, "Exploiting programmable temperature compensation devices to manage temperature-induced delay uncertainty," *IEEE Trans. Circuits and Systems I*, vol. 59, no. 4, pp. 735–748, Apr. 2012.

3. Integrated Circuits (Microcircuits) Manufacturing, General Specification, Std. MIL-PRF-38535H, U.S. Dept. of Defense, Mar. 2007.

4. T. Ragheb, A. Ricketts, M. Mondal, S. Kirolos, G. M. Links, V. Narayanan, and Y. Massoud, "Design of thermally robust clock trees using dynamically adaptive clock buffers," *IEEE Trans. Circuits and Systems I*, vol. 56, no. 2, pp. 374–383, Feb. 2009.

5. T. Sakurai, "Alpha-power law MOSFET model and its applications to CMOS inverter delay and other formulas," *IEEE Trans. Solid-State Circuits*, vol. 25, no. 2, pp. 584–594. Apr. 1990.

6. A. Bellaouar, A. Fridi, M. J. Elmasry, K. Itoh, "Supply voltage scaling for temperature insensitive CMOS circuit operation", *IEEE Trans. Circuits Syst. II Express Briefs*, vol. 45, no. 3, pp. 415–417, Mar. 1998.

7. R. Kumar and V. Kursun, "Reversed temperature-dependent propagation delay characteristics in nanometer CMOS circuits", *IEEE Trans. on Circuits and Systems*, vol. 53, no. 10, pp. 1078–1082, Oct. 2006.

8. A. Calimera, E. Macii, M. Poncino, and R. I. Bahar, "Temperature insensitive synthesis using multi-vt libraries," in *Proc. 18th ACM Great Lakes Symp. VLSI*, May 2008, pp. 5–10.

9. Predictive Technology Model (PTM), http://www.eas.asu.edu/~ptm.

10. M. L. Alles, L. W. Massengill, S. E. Kerns, K. L. Jones, J. E. Clark, and W. F. Kraus, "Effect of temperature-dependant bipolar gain distribution on SEU vulnerability of SOI CMOS SRAMS," in *Proc. IEEE 96*, 1992.

11. H. I. T. Ooka, "Temperature dependence of single-event charge collection in SOI MOSFET's by simulation approach," *IEEE Trans. Electron Devices*, vol. E80-C, no. 3, p. 417, Mar. 1997.

12. O. A. Amusan, A. F. Witulski, L. W. Massengill, B. L. Bhuva, P. R. Fleming, M. L. Alles, A. L. Sternberg, J. D. Black, and R. D. Schrimpf, "Charge collection and charge sharing in a 130 nm CMOS technology," *IEEE Trans. Nucl. Sci.*, vol. 53, pp. 3253–3258, Dec. 2006.

13. S. DasGupta, "Trends in Single Event Pulse Widths and Pulse Shapes in Deep SubMicron CMOS," M.S. thesis, Vanderbilt Univ., Dept. Electrical. Eng., Tennessee, 2007.

14. G. Guo, T. Hirao, J. S. Laird, S. Onoda, T. Wakasa, T. Yamakawa, and T. Kamiya, "Temperature dependence of single event transient current by heavy ion microbeam on p+/n/n+ epilayer junctions," *IEEE Trans. Nucl. Sci.*, vol. 51, pp. 2834–2839, Oct. 2004.

15. P. E. Dodd and L. W. Massengill, "Basic mechanisms and modeling of single-event upset in digital microelectronics," *IEEE Trans. Nucl. Sci.*, vol. 50, pp. 583–602, Jun. 2003.

16. C. Shuming, L. Bin, L. Biwei, and L. Zheng, "Temperature dependence of digital SET pulse width in bulk and SOI technologies," *IEEE Trans. Nucl. Sci.*, vol. 55, no. 6, pp. 2914–2920, Dec. 2008.

17. A. Balasubramanian et al., "Crosstalk effects caused by single event hits in deep sub-micron CMOS technologies", *IEEE Trans. on Nuclear Science*, vol. 53, no. 6, Dec. 2006.

18. W. Zeng, D. Gang, Y. Yintang and L. Jianwei, "Crosstalk noise voltage of coupling RC interconnects with temperature distribution," *Chinese Journal of Electronics*, vol. 19, no. 1, 2010.

19. S. Sayil, V.K. Boorla, S.R. Yeddula, "Modeling Single Event Crosstalk in Nanometer Technologies", *IEEE Trans. on Nucl. Sci.*, vol. 57, no. 5, pt. 2, pp. 2493–2502, October 2011.

20. A. Hokazono et al. "MOSFET design for forward body biasing scheme", *IEEE Electron Device Letters*, vol. 27, no. 5, pp. 387–389, 2006.
21. I. M. Filanovsky and S. T. Lim. "Temperature sensor applications of diode-connected MOS transistors," in *Proc. IEEE Int. Symp. Circuits and Syst.*, May, 2002, pp. II-149.

PVT Variations

8

8.1 Introduction

Designing digital circuits in nanoscale CMOS technology faces many challenges in terms of variations. There are three major sources of variations: Process, Voltage and Temperature variation. These variations are altogether named and referred as PVT variations. Variations make some parts of the chip slower than others, either because the transistors are intrinsically slower or because of a high temperature or low supply voltage. Thus, circuits in these sections may not be able to propagate signals fast enough and may suffer timing errors [1].

In an ideal design, supply voltage is supposed to be constant irrespective of the location. Power is distributed to all transistors on the chip with the help of a power grid network. When a large number of gates switch simultaneously in an IC, a significant amount of current is drawn from the power distribution network and then returns to the ground distribution network. Due to the resistance of the distribution lines, resistive voltage drops occur on power and ground lines. The voltages on supply and ground wires are no longer constant and include noise component which include DC and high frequency inductive noise.

IR drop is a dynamic event that occurs due to simultaneous switching events such as clocks, large buffers and drivers. The impact will be worsened by the presence of Ldi/dt voltage variations occurring at package pins. Ldi/dt voltage drops are usually referred as the high frequency noise component of the power supply noise. While the package pin inductance remained relatively constant over the years, increasing circuit frequency is responsible for increasing Ldi/dt voltage drops occurring in power grid.

Due to resistive drops and inductive noise in the power supply, the supply voltage can vary causing dramatic effects on circuit performance. As the power supply voltage and ground voltage level are changed due to power supply noise, the transistor currents will

© The Author(s), under exclusive license to Springer Nature Switzerland AG 2022   131
S. Sayil, *Noise Contamination in Nanoscale VLSI Circuits*, Synthesis Lectures on Digital
Circuits & Systems, https://doi.org/10.1007/978-3-031-12751-9_8

be impacted. It impacts propagation delay of logic gates since current drive capabilities of devices are affected. In other words, the propagation delay of logic gates strongly depends on the power supply voltage level. Power supply noise can degrade noise margin of logic gates impacting overall timing and circuit functionality [2].

On-chip temperature gradient has also emerged as a major design concern for high-performance integrated circuits for nanoscale technology nodes as it significantly affects the performance of a chip. Some parts of the chip can be more densely packed and could have more active switching compared to the other parts of the chip. In these regions, increased temperature effect will be seen and in turn would cause large temperature gradients. On-chip temperature gradient can be as high as 50 °C.

The increased temperature affects both interconnect resistance and driving strength of buffers. Since the driving strengths of the clock buffers are impacted along with interconnect, clock skew phenomenon for synchronous digital circuits is further exacerbated by the thermal gradients [3]. In addition, increasing circuit temperature deteriorates circuit reliability and increases circuit sensitivity to soft error and crosstalk events.

The resistance of an interconnect increases linearly with temperature due to positive temperature co-efficient of materials used such as copper. The impact of temperature on a driver is complicated due to the fact that carrier mobility and device threshold voltage has conflicting effects on the current when temperature changes. As temperature increases, the carrier mobility decreases which causes the device current to reduce. With increased temperature, device threshold on the other hand reduces causing drive current to increase. Depending on the supply voltage level, device current may increase or decrease. At nominal supply voltage around 1 V, change in carrier mobility dominates over the change of threshold voltage which causes the device current to decrease with increased temperature.

Process variations are caused by the inability to precisely control the fabrication process at nanoscale technologies. Systematic errors such as non-ideal properties in lithographic process and random errors like changes in dopant density [4]. Process variation is defined as the difference between the intended and the actual transistor where the difference is caused by changes in fabrication process equipment, material, and processing.

Variations in the process parameters, such as impurity concentration, oxide thicknesses, and diffusion depths, caused by non-uniform conditions during the deposition and/or the diffusion of the impurities.

Due to limited resolution of the photolithographic process, changes in transistor dimensions such as widths and lengths may occur giving different (W/L) ratios of MOS transistors. On the other hand, changes in oxide thickness, substrate, poly and implant impurity levels can vary the threshold voltage of a device.

In order to aid designers, the manufacturers commonly specify fast and slow device corners in addition to the nominal operation condition.

There are two categories of variations: Namely, inter-die and intra-die variations. Inter-Die variation is defined as variations occur from one die to another, but properties remain almost same in one die. In the case of Intra-Die or within-die variations, a device feature varies between different locations on the same die. This is also known as local mismatch. Intra-die variations are increasingly becoming an important factor in semiconductor chips.

Intra-die variations can create functional failures even in non-worst case corners as they affect the transistors differently. Intra-die variations create a difference in electrical properties of neighboring devices. It can increase skew between two clock paths.

Systematic variations are variations in fabrication process due to non-uniform equipment. The impact of these variations on performance can be predicted using simulation models and design or equipment data. Their effect can then be minimized by modifying the fabrication process.

Random variations can be defined as statistical fluctuations in manufacturing equipment introducing performance variations between different dies or between different elements in the same die. Random variations are not predictable.

Among the various components of a synchronous design, clock networks are most sensitive to intra-die variations owing to their differential nature, namely two parameters: pulse-width and skew.

Variations arise from fluctuations in various steps of fabrication process such as deposition, etching, sub-wavelength lithography, and Chemical Mechanical Polishing.

Photolithography refers to the process of using light to transfer a pattern on a photomask to a silicon wafer coated with light sensitive material (photoresist). Etching and photolithography creates the source of most variations. Doping is required to create wells, channel, gate, drain and source junctions. Random variations in dopant locations inside the devices can also create variations. Channel doping has the most impact on transistor characteristics and any variation in average doping level can create large variations in the current and threshold voltage.

Chemical Mechanical Polishing or CMP is the process to flatten out the deposited layers and reduces the layers to desired thickness. However, it may be source of random variation. Interconnect cross-section consists of metal width and thickness. Metal width variations happen due to the etching and lithography variations and metal thickness variations arise from CMP and Inter Layer Dielectric (ILD) variations [5].

ILD thickness and permittivity variations affect metal thickness, current and coupling capacitance in metal lines [6]. Gate length or width variation arises from lithography and etching fluctuations.

References

1. S. Bhunia, S. Mukhopadhyay, K. Roy, "Process Variations and Process-Tolerant Design," *Proc. of Intl. Conf. on VLSI Design*, 2007, pp. 699–704.
2. T. Ragheb, A. Ricketts, M. Mondal, S. Kirolos, G. M. Links, V. Narayanan, and Y. Massoud, "Design of thermally robust clock trees using dynamically adaptive clock buffers," *IEEE Trans. Circuits and Systems I*, vol. 56, no. 2, pp. 374–383, Feb. 2009.

3. J.M Rabaey, A.P. Chandrakasan, B. Nikolić, Digital integrated circuits: a design perspective, 2nd edition, Pearson Education, 2003.
4. S. Nassif, K. Bernstein, D.J. Frank, A. Gattiker, W. Haensch, B.L. Ji, E. Nowak, D. Pearson, N.J. Rohrer, "High Performance CMOS Variability in the 65 nm Regime and Beyond," *IBM Journal of Research and Development*, Vol. 50, Issue 4/5, pp. 433–449, July 2006.
5. N. Verghese, P. Hurat, "DFM reality in sub-nanometer IC design," *ASP-DAC'07*, Jan. 2007, pp. 226–231.
6. M. Anis, M.H. Aburahma, "Leakage current variability in nanometer technologies," *Proc. 5th Intl. Workshop on SoC for Real-Time App.*, July 2005, pp. 60–63.

Index

Printed in the United States
by Baker & Taylor Publisher Services